CALIFORNIA STATE UNIVERSITY, HAYWARD
LIBRARY

World Resources

Gunnar Alexandersson
Björn-Ivar Klevebring

World Resources

Energy · Metals · Minerals

Studies in Economic
and Political Geography

Walter de Gruyter · Berlin · New York 1978

GeoSpectrum
Editor of the series: Dr. Wolf Tietze
                         D-3180 Wolfsburg

Authors

Gunnar Alexandersson
Professor of International Economic Geography
The Stockholm School of Economics

Dr. Björn-Ivar Klevebring
Technologist and Metallurgist
Head, Metallverken Pyrometallurgical Research

CIP-Kurztitelaufnahme der Deutschen Bibliothek

Alexandersson, Gunnar
World resources : energy and minerals ; studies in economic and polit. geography / Gunnar Alexandersson ; Björn-Ivar Klevebring. – 1. Aufl. – Berlin, New York : de Gruyter, 1978.
(GeoSpectrum)
ISBN 3-11-006577-0
NE: Klevebring, Björn-Ivar:

Library of Congress Cataloging in Publication Data

Alexandersson, Gunnar.
　World resources, energy and minerals.
　(GeoSpectrum)
　Bibliography: p.
　　1. Natural resources. 2. Power resources. 3. Mineral industries. 4. Geography, Economic. 5. Geography, Political. I. Klevebring, Bjorn-Ivar, 1943- joint author. II. Title. III. Series.
HC55.A36   333.7   77-27560

© Copyright 1978 by Walter de Gruyter & Co., Berlin. – All rights reserved, including those of translation into foreign languages. No part of this book may be reproduced in any form – by photoprint, microfilm, or any other means – nor transmitted nor translated into a machine language without written permission from the publisher.
Cover design: T. Bonnie. – Printing: Walter de Gruyter, Berlin. – Binding: Lüderitz & Bauer, Berlin. – Printed in Germany.

Preface

This book deals with a functionally well-defined problem: the exploitation and utilization of the products of the mineral realm. The word mineral refers to substances that are obtained from the earth's crust. The word has also a more narrow meaning which is relevant in the book: naturally occurring elements or substances with a definite chemical composition. It should be clear from the context if the wider or the more narrow meaning applies.

The exploitation of minerals from the crust of the earth and the utilization of plants and animals on the surface of the earth (agriculture, forestry, collection of berries, nuts and so on, animal husbandry, hunting and fishing) have always been central topics of research in economic geography. Textbooks have traditionally devoted much space to raw materials, the physical basis of our civilization.

Mineral exploitation is a dynamic industry, full of drama. New products based on mineral raw materials are continuously being developed in the research laboratories of industrial countries. Demand for a new mineral or an increase in the demand for an old one can suddenly give a new lease on life to a depopulated region or create new jobs in an underdeveloped country. The oil boom in the Persian Gulf area, which in a few decades carried some of the poorest lands anywhere to the top of a ranking list according to average gross national product per capita, may serve as the extreme case. Some poor countries won the highest prize ever on the raw material lottery.

In spite of the dynamism of mineral exploitation, which was strongly emphasized in Erich W. Zimmermann's classic World Resources and Industries, published in New York in its second edition 1951, the geography of raw materials was moved into the shadow on curriculums of geography departments in Western Europe and North America in the 1960s. The old standard bearer of economic geography was pushed to the side for courses in the theory of economic geography, concerned primarily with retail trade and other tertiary activities.

People directly involved in raw material production make up a small percentage in industrial countries, but raw materials still play an important part in interregional and international trade. Raw material production is a base industry of great direct importance to large segments of the secondary and tertiary sectors of the economy. These segments are the manufacturing and commercial superstructure of the primary industry. No wonder, raw materials still play a prominent part in the business pages of our newspapers and financial journals.

In the 1970s raw materials came into the focus of world politics as never before. The assumption of classic economic theory, that omniscient man lives in a global country and uses first his richest soil and mineral resources and from there proceeds to poorer and poorer resources, creating a land rent in the center, was proved to be fundamentally unrealistic. Far from being omniscient, inventive man started agriculture and mineral production where he happened to live, as a rule in rather poorly endowed regions. Europe did not have superb coal deposits, although they were large. United States was not uniquely endowed with petroleum resources, although they were significant. Production and transportation costs created the high land rents not in the center but in the periphery. Instead of harmony and balance provided by the classic economic theory, the real world is plagued by serious imbalances, heightened by the fact that the world is not one but made up of some 150 countries, most of which may want to assert their independence.

The center and the periphery as a rule are located in separate countries. Sparsely populated nations in the periphery, in possession of easily obtained mineral resources much needed in the densely populated industrial center, may withhold production and expose the center to monopolistic pricing with sudden changes in the international flow of capital and goods and serious balance-of-payment-adjustments as a result.

The center with its superior technology and scientific capacity is not without options. Substituting one type of energy for another, one metal for another, is a major alternative. Utilizing the minerals of the sea bed is also an alternative: the continental shelf (petroleum, diamonds, gold, tin and others), the continental rise (petroleum), the deep sea floor (the manganese nodules: manganese, nickel, copper and cobalt), or the mid-ocean rift valley. The nations of the industrial center are therefore anxious to establish regimes for the exploitation of the mineral resources of the sea. Countries of the periphery, forming a solid majority of votes in the United Nations, see this alternative in a different light.

It is our hope that this survey of an important part of the raw-material sector will serve as an orientation for students of high schools and colleges and not the least for the newspaper reading public. This book does not use the long perspective common in similar texts of recent years written in a more philosophic vein. Its focus is the present, with the past hundred years a background and the next two or three decades as the time span for implicit or explicit projections. We try to strike a balance between the *après-nous-le-deluge*-philosophy and the exponential growth scare. A close look at the recent past and a probable near future show that mankind is past the inflection point on a logistic curve both for population growth (not treated here) and for consumption of most mineral resources.

One mineral, the most important of them all, was left out of the book. Fresh water and air were long seen as free goods and therefore of little interest to economists who study the husbanding of scarce resources. However, water in recent times has been reclassified and the water supply in many places has become an important task for the geologists. The price of water is usually so

Contents

Preface V
Production and Reserves 5
    Exponential Growth or the Logistic Curve? 11
Energy 15
Fossil Fuels 19
    Coal 19
        Origin of coal 19, Types of coal 20, Occurrence of coal 21
        Europe 23
            United Kingdom 23, Continental Western Europe 26, Central Europe 28, Other Europe 30
        North America 30
            United States 30, Canada 34
        USSR 35
            Donbass 35, Kuzbass 35, Other fields 36
        Asia 36
            China 36, Japan 37, Other Asia 38
        Southern Hemisphere 38
            Australia 38, South Africa 39, Rhodesia 40, Latin America 40
        Outlook for the Coal Industry 40
    Petroleum – Oil and Natural Gas 41
        Oil 43
            Oil refining 45, Transportation 46, Location of oil refineries 46
        North America 47
            United States 47, Canada 49, Mexico 50
        South America 52
            Trinidad 52, Venezuela 52, Colombia 54, Peru 55, Ecuador 55, Argentina 55
        Europe 55
            Romania 56, Germany 56, Norway 56, United Kingdom 56
        The Soviet Union 57
            Baku 57, The North Caucasus fields 57, The Volga-Urals-region 58, Northwestern Siberia 58, Scattered fields 60, Pipelines 60, Prospects 61
        The Gulf 61
            Iran 61, Iraq 62, Bahrain 63, Saudi Arabia 63, Kuwait 64, The United Arab Emirates 65, Company-State relations 65, Oil and Geopolitics 65
        Africa 66
            Algeria 66, Libya 66, Nigeria 67, Other Africa 67
        Other Asia 67
            Indonesia 67, China 68, India 69, Burma 69
        Oceania 69
            Australia 69
        Prospects of the Oil Industry 70
    Natural Gas 70
        North America 73
            United States 73, Canada 75
        Western Europe 75
        Soviet Union 77

Contents

Nuclear Power	80
History 80, The atom 81	
Fission Energy	82
Light Water Reactors	83
Enrichment plants 85, Fuel recovery plants 86, The future of nuclear power 86, Great Britain 87, Germany (W) 89, France 89, United States 90, Sweden 90, USSR 91, India 91	
Breeder Reactors	92
Fusion Energy	93
Water Power	94
Hydroelectric power stations 95	
Electric Energy	97
Waste heat 103, Radioactivity 104, Other environmental deterioration 104	
Metals and Other Minerals	107
Geology	107
Iron Ore	108
Ore types 108, Iron and steel making 109	
North America	111
United States 111, Canada 114	
Latin America	115
Brazil 115, Venezuela 115, Chile 115, Peru 116, Cuba 116	
Western Europe	116
Sweden 116, Norway 117, Finland 117, France 117, Germany 117, United Kingdom 118, Austria 119, Spain 120	
Central Europe	120
Yugoslavia 120	
USSR	120
The Ukraine 120, Kursk Magnetic Anomaly 122, The Urals 122, Scattered Fields 123	
Asia	123
Japan 123, China 123, India 123, Other Asia 124	
Australia	125
Africa	125
South Africa 125, Liberia 126, Mauritania 126, Central African Republic 126	
Alloy Metals	127
Manganese	127
Chromium	129
Nickel	129
Tungsten	131
Antimony	132
Molybdenum	133
Vanadium	134
Cobalt	134
Columbium	135
Tantalum	135
Rare Earth Metals	135
Base Metals	136
Copper	137
North America	138
United States 138, Canada 140	

South America	140
Chile 140, Peru 142	
Africa	142
Zaire 142, Zambia 143	
USSR	143
Mongolia 145	
Australia	145
Europe	145
Tin	146
Southeast Asia 147, Bolivia 147, Africa 149, USSR 149	
Zinc	149
United States 150, Canada 150, Australia 150, Other producing regions 152	
Lead	153
United States 154, Canada 154, Mexico 154, Australia 154, USSR 156, Europe 156	
Mercury	156
Italy and Spain 156, USSR 156, North America 156, Pollution of the environment 156	
Precious Metals	157
Gold	158
South Africa 159, USSR 160, Canada 160, United States 161	
Silver	161
The Western Hemisphere 162, USSR 162	
Platinum	162
Light Metals	163
Aluminum	163
History 163, Bauxite 164, Alumina 164, Aluminum 165, Economics 165, Uses 165, Corporations 166	
Europe	166
USSR	168
North America	170
Latin America	171
Africa and Asia	174
Australia	174
Magnesium	175
Titanium	176
Nonmetallic Minerals	177
Stone	178
Cement	179
Technique 180, Location 181, The cement industry and economic theory 183	
Asbestos	184
Clay	185
Sand and Gravel	185
Glass	185
The flat-glass industry 186, Glass containers 187, Special glass 187	
Porcelain	188
Ball clay	189
Fire clay	189
Bentonite	189
Fuller's earth	190

Contents

Diatomite	190
Graphite	190
Sulfur	190
Fertilizers	192
Nitrogen	194
Chile saltpeter 195, Norwegian saltpeter 196, Ammonia 196	
Phosphorus	197
United States 198, North Africa 199, Soviet Union 200, Asia 200, Oceania 200, Latin America 200	
Potash	201
Europe 201, North America 202, USSR 202	
Calcium	203
Abrasives	203
Industrial Diamonds	203
Aluminum Oxide and Other Natural Abrasives	205
Corundum 205, Emery 205, Garnet 205, Tripoli 205, Rottenstone 205	
Boron	205
Mineral Raw Materials in International Trade	207
Raw materials and political geography 208, Petroleum and other minerals 210, International trade in minerals 213	
Notes	219
Measure Prefixes	225
Capacity	225
Energy	225
Short Bibliography	227
Statistical Sources	229
Index	231

Production and Reserves

World production of a metal or a fossil fuel is often compared with the global reserves of the mineral. This tells how long the reserves will last. But both quantities in the equation are very approximate and geologists and other experts are cautious with their statements. For obvious reasons, reserves are not known but at best approximately estimated. With technical progress, resources that are presently not exploitable can be turned into reserves. On the other side of the equation, the future demand is exceedingly difficult to forecast.

The Earth is a limited planet, whose dimensions have been approximately known since the beginning of the 16th century and whose form and size were hypothetically debated in ancient Greece. The produced quantities of minerals, however, are tiny in relation to the volume of the Earth or its crust. The statement that the Earth is limited does not carry the argument very far.

Exploitation of mineral resources on or in the seabed, that covers 71 percent of the surface of the Earth, has just started. Large scale extraction of minerals from sea water is also relatively new. From the human point of view the sea is tremendous but by the scale of the Earth it is just a patch of moisture. On a globe with a diameter of 3 meters the average depth of the sea (3 800 m) should be represented by 1 millimeter.

Before 1960 mineral production in the sea was unimportant. Only three or four countries obtained some petroleum from wells on the continental shelf off their shores. The Convention of the Continental Shelf, adopted by the UN Law of the Sea Conference at Geneva 1958, went into effect 1964 after ratification by the required number of states. The Convention provides a legal definition of the shelf concept − it extends out to a depth of 200 m − which roughly coincides with the old geological definition. However, recent measurements indicate that the shelf edge on the average is at the 132 m isobath and that it varies between less than 70 m and more than 600 m. In the Convention text the coastal state is confirmed in its right to exploit mineral resources on and in the seabed of the legal shelf and beyond to depths admitting exploitation. Although the first part of the rule is simple and helped get the boom in offshore oil drilling under way, the expansible last part, inserted to satisfy South American nations, would make a new sea conference necessary as soon as the technique of deep sea mining had developed to the point where large-scale exploitation of the manganese nodules on the deep sea floor, petroleum from the continental rise or various metals from the rift area became technically possible and economically feasible. The conference meeting in 1974, 1975 and 1976 has had its attention primarily focused on the nodules.

In 1969 oil prospecting was carried on off the coasts of 75 countries in all continents except Antarctica. The technique was yet in its initial stage, but, still, off-shore oil was obtained in over thirty countries. Off-shore fields in 1969 accounted for 17−19 percent of the oil production in the world and for over six percent of the natural gas.

These shares were expected to increase substantially in the following decade. Offshore exploitation of diamonds, gold, tin and other metals on the shelf were also on their way. Since these are heavy substances they are primarily found under shallow water near the shore. Avoiding the surf seems to be a greater problem for this type of mining than working under a high water column. Several vessels have been lost off the coast of South Africa recovering diamonds.

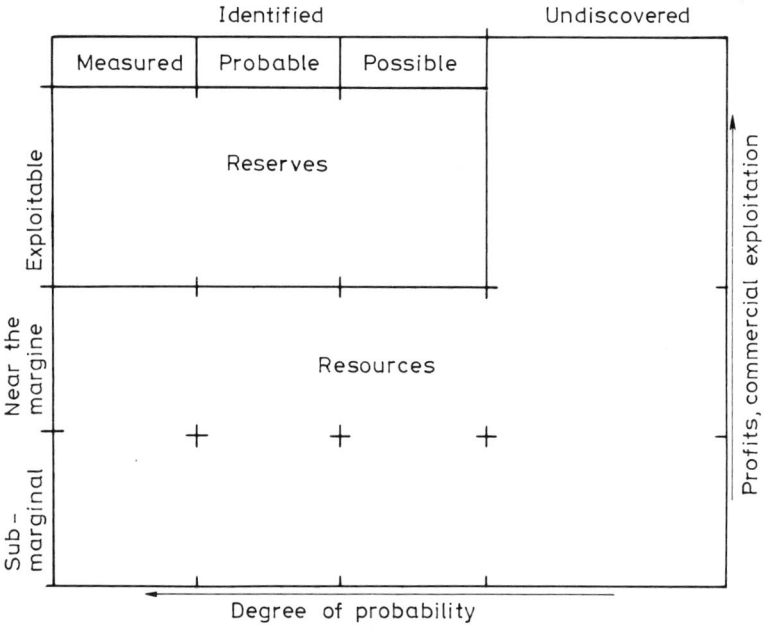

Fig. 1 − Classification of mineral deposits into reserves and resources. The degree of security increases from right to left in the diagram. Proven reserves are at the left. Profitability increases from bottom to top. The highest land rent is at the top of the figure.

A distinction is made between reserves, which are deposits exploitable with present techniques, known resources, which are reserves plus not yet economically exploitable deposits, and total resources, which are known resources plus not yet discovered resources. The border line between exploitable and non-exploitable deposits shifts with the economic and with the technical development.

A temporary shortage of a mineral is overcome by price increases. A shortage of reserves is corrected by a policy that accelerates prospecting, develops new prospecting techniques and furthers metallurgical research into new process methods, which may allow the use of ores with a lower metal content or a different type of ore, e. g. lateritic oxide ore which globally contains much more nickel than the conventional sulfide ores.

Source: V. E. McKelvey, "Mineral Potential of the United States", in The Mineral Positions of the United States 1975−2000, E. N. Cameroon, ed. (Madison: University of Wisconsin Press, 1973).

Mineral deposits on land are also incompletely mapped. For example, the known reserves of iron, the most important and best researched metal of our civilization, more than doubled in the 1960s. The prospecting rush in the Pilbara district of Australia alone revealed more iron ore of high metal content than the reserves listed for the world in 1960. Ore is a mineral deposit that can be exploited with present technique. In 1700, copper ore holding 13 percent metal could be utilized. In 1900, the limit had been lowered to 2.5 percent and now to less than 0.5 percent. Ore in itself is an expansive concept.

In developed countries (DCs) with large territory in relation to their population, like the Soviet Union, Canada and Australia, it has been possible only after the Second World War to survey almost their whole territory thanks to airborne magnetometers and other advanced techniques. New mineral finds have been made in rapid succession in such areas. Newly rich oil countries can be expected to prospect for other mineral deposits than oil and natural gas in the near future. Even in DCs it usually takes a long time before all financial, metallurgical, logistical and marketing problems have been solved and the project can proceed from mapping and analysis to exploitation. Many interests have to be brought to cooperate. In a country with a market economy it is not only the mining company with its bank, customers and employees that are

Tab. 1 — World Reserves and Resources of Selected Minerals: 1948 and 1972
  (Million tons if not otherwise stated)

Mineral	1948 reserves	1972 reserves	1974 known resources
Bauxite	1 400	5 900	9 800
Chromium	100	700	2 800
Copper	100	308	no data
Iron ore*	19 000	251 000	782 500
Lead	40	75	no data
Manganese	1 000	3 900	15 200
Mercury (flasks · 10^3)	no data	7 000	15 000
Silver (kg · 10^6)	85.54	171.1	634.5
Tin	6 000	5 700	11 600
Titanium	no data	520	1 810
Tungsten (ton · 10^3)	4 000	1 360	no data
Uranium (ton · 10^3)	no data	944	2 500
Vanadium (ton · 10^3)	no data	4 200	18 100
Zinc	70	124	no data

* The iron ores of the USSR were not included in the 1948 reserves; they accounted for 44 percent of the 1972 reserves. The new reserves of the Pilbara region in Australia were not included in the 1972 UN estimates. The statement in the text was based on local Australian estimates from 1970.

Source: G. J. S. & M. H. Govett, "The Concept and Measurement of Mineral Reserves and Resources," Resources Policy, Vol. 1, 1974, p. 46—55.

8   Production and Reserves

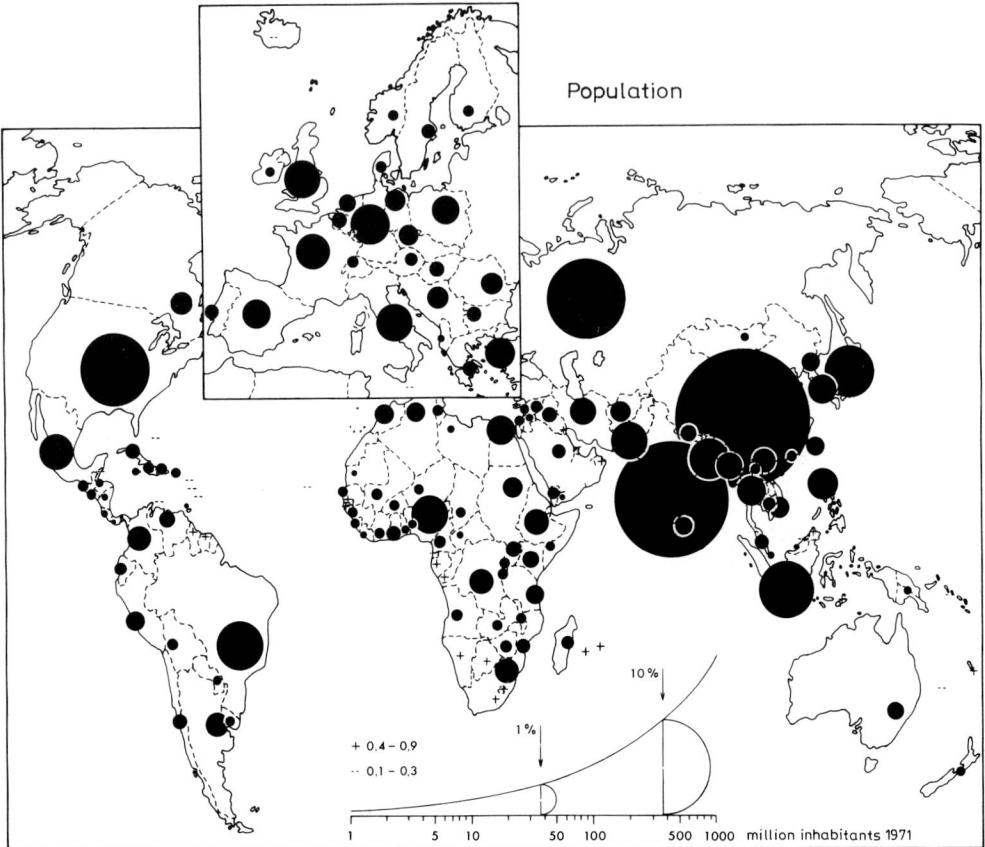

Fig. 2 – Among the DCs, the United Nations includes Anglo-America (Canada and the United States), Europe including the Soviet Union, Japan, Australia, New Zealand, and South Africa. Other countries are LDCs.

interested in the mine but also authorities on various levels: federal, state and county. In recent years people in towns, regions and the country have made their views known through demonstrations and meetings. With increasing scale of the mining operation and related transportation, ecological and esthetical points of view must be weighed into the economic balance sheet. This at least holds for the richest parts of the rich regions, e. g. the New England and Middle Atlantic coasts of the United States. The delays in the construction of the trans-Alaskan pipeline and in the opening of the Montana-Wyoming coal fields show that environmental pressure groups have a say also in peripheral parts of industrial countries.

Mining mineral deposits in a foreign country is likely to be more risky than doing it in the home country. It is most hazardous if the host country is underdeveloped and lacks infrastructure, if it is poor and politically instable.

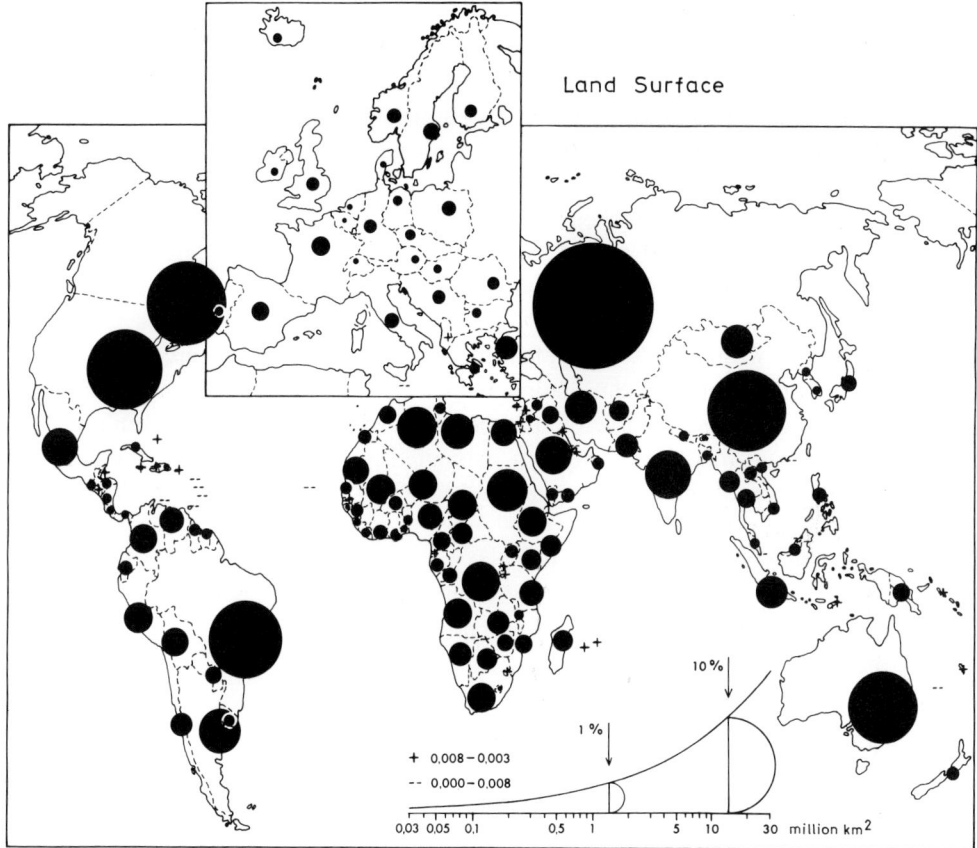

Fig. 3 – The symbol scales of the two base maps, figs 2 and 3, are chosen to facilitate direct comparison. The sum of the symbol surfaces on the two maps – and on a series of maps in the book – is the same. For example, 0.1%, 1.0% or 10% of the world total are shown with three symbols of equal size on all maps, see symbol scales. This makes possible a direct comparison of quantities expressed in US dollars, tons, kWh, inhabitants or km². For a third base map – showing standards of living – electricity production was chosen, fig 22. In this book that indicator also has an interest of its own.

Trade in minerals therefore is primarily trade between industrial countries. Canada plays a greater role for the mineral supply of the United States than all of Latin America and Africa combined, figure 42. Australia is more important for Japan than all of Southeast Asia, figure 43.

For less developed countries (LDCs) in Latin America, Africa and Asia the mineral trade that often plays a prominent part on the export side of their trade balance usually is only the cream of their total mineral reserves. If copper ore is being mined it will hold 2–6 percent metal (Zambia 3.4 percent) while American multinational corporations domestically mine ore with an average of 0.6 percent copper. In the LDCs most mineral deposits have not yet been discovered. Systematic prospecting has never occurred. If the LDCs are going to find and exploit these deposits themselves, they have a long way to go. It takes much time and capital to train geologists and business managers, to construct mines, infrastructure and smelters and not the least to develop the sophisticated industry that is going to consume the minerals. The tremendous increase in population in the next few decades will force the LDCs to consume an increasing share of their rapidly increasing mineral production.[1] But they need technology and capital from the DCs to find and develop their mineral resources. They need the income of the mineral exports to finance their general economic development.

Groups within the DCs that fear that their own mineral resources will soon be exhausted should find consolation in the large territory of all the DCs combined. They have also incompletely prospected areas, many undetected mineral resources and many conservatively estimated mineral reserves. In addition, the exploitation of the enormous mineral deposits of the seabed has only recently started on a small scale.[2]

Trade between industrial countries will contribute to the necessary consumption needs even in the future. New energy sources will permit exploitation of ores with lower and lower grade and production of substitutes for metals that are being exhausted.

Tab. 2 — Composition of the Earth's Crust, the Lithosphere

Order	Element	Weight percent	Order	Element	Weight percent
1	Oxygen (O)	46.6	11	Phosphorus (P)	0.12
2	Silicon (Si)	27.7	12	Manganese (Mn)	0.10
3	Aluminium (Al)	8.1	13	Fluorine (F)	0.07
4	Iron (Fe)	5.0	14	Sulfur (S)	0.05
5	Calcium (Ca)	3.6	15	Strontium (Sr)	0.05
6	Sodium (Na)	2.8	16	Barium (Ba)	0.04
7	Potassium (K)	2.6	17	Carbon (C)	0.03
8	Magnesium (Mg)	2.1	18	Chlorine (Cl)	0.02
9	Titanium (Ti)	0.4	19	Chromium (Cr)	0.02
10	Hydrogen (H)	0.1	20	Zirconium (Zr)	0.02
Total of the first 10		99.2	21	Rubidium (Rb)	0.01
			22	Vanadium (V)	0.01
			23	Nickel (Ni)	0.01
			24	Zinc (Zn)	0.01
			25	Nitrogen (N)	0.01
			Total of above 25		99.73

Source: Raymond L. Parker, *Data of Geochemistry* (Washington: GPO, 1967, Geological Survey, Prof. Paper 440−D).

The lithosphere or the firm crust (terra firma) to 95 percent is made up of igneous rocks (including metamorphic rocks) and only to 5 percent of sedimentary rocks (4 percent shale, 0.75 percent sandstone and 0.25 percent limestone). The composition of the lithosphere therefore largely coincides with that of the igneous rocks.

Tab. 3 — Physical Dimensions of the Earth

Layer		Thickness (km)	Density	Mass (%)
Atmosphere		–	–	0.00009
Hydrosphere	Earth's crust	3.8 (average)	1.03	0.024
Lithosphere		17 (average)	2.8	0.4
Mantel		2883	4.5	68.1
Core		3471	10.7	31.5
Earth, total		6371 (average)	5.52	100.0

Source: Brian Mason, *Principles of Geochemistry* (New York: Wiley, 1958, 2nd ed.) p 41.

The continents are estimated to be made up of $105 \cdot 10^6$ km² continental shields, on the average covered by 500 m of sediments, and of $42 \cdot 10^6$ km² young fault zones with an average of 5 km of sediments. The deep sea sediments are on the average 300 m and the shelf sediments ($30 \cdot 10^6$ km²) 4 km. Sediments on the shelves are similar to those of the young fault zones.

Exponential Growth or the Logistic Curve?

In the first edition of his anonymously published An Essay on the Principle of Population 1798 Thomas Malthus writes about the arithmetic progression at which food production grows and the geometric progression at which the human population increases. This mathematic statement was left out in later editions but was to play an important role in the ensuing discussion, not the least in modern 'doomsday prophecies'.

Malthus met much opposition already in the 19th century. The geometric progression, or as it is now usually called, the exponential growth after a historically short period leads to absurd results. The American economist Henry George in his Progress and Poverty jokingly suggested that Adam, using the same type of mathematics for projecting the future weight of his first child, would have found, assuming it weighed 4 kg at birth and 8 kg at the age of eight months, that at thirty years of age it would weigh 140 billion tons.

12 Exponential Growth or the Logistic Curve?

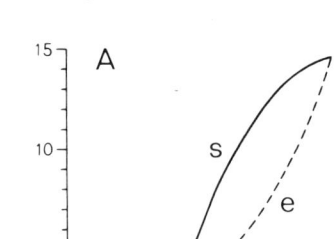
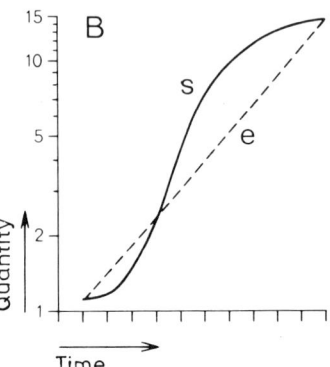

Fig. 4 – The diagrams show development between two dates: with exponential growth (e) and with growth along an 'S-curve' (s); with the quantity axis in arithmetic scale (A) and in logarithmic scale (B). In the former, (e) is a 'J-curve', in the latter, a straight line.

The inflection point, where the S-curve shifts from being concave to being convex, is difficult to recognize in reality. Projections based on exponential growth may be reasonably good for a time-span around the inflection point. For long periods they are absurd.

Exponential growth has a certain doubling time. With an increase of 7.2 percent a year, for instance, the doubling time will be 10 years. Exponential growth will be most dramatic seen in hindsight. Seen from the point when a given resource is being exhausted, 50 percent were left in the ground at the beginning of the last doubling time (t). At the beginning of the last but one, 2 t from now, 75 percent were left and 3 t from now 87.5 percent. Some 4 t ago 93.75 percent were left and 5 t ago 96.875 percent. If the exponential growth was high the resource was almost untouched not long ago. Already at an early stage the 'doomsday prophet' starts to get nightmares about the day of the Last Judgment soon to come. The belief in exponential growth is one thing all doomsday prophets have in common.

The statistician Quetelet and the mathematician Verhulst in the 1830s suggested the 'S-shaped' or logistic curve as more probable than the geometric curve of Malthus for describing the population development. In order to be logically sound the logistic curve should represent expansion within a closed system in which the curve threatens to hit the ceiling. The Earth is such a closed system. The calculation of how long it would take with the population increase of 1970 (2.0 percent a year) for the population of the world to be crammed with standing room only or to represent a mass equalling the mass of the earth may be an interesting exercise in applied mathematics, but it lacks all interest for human planning. Humanity in all likelihood will adapt itself through growth rate zero or a population decline and through a limit on the production of non-renewable materials or a substitution of renewable for non-renewable raw materials.

The population of the DCs have almost completed the demographic transition from a low growth rate due to high birth rates and high death rates, which characterized the period before the industrial revolution, to a low growth rate due to low birth and death rates, which are a prerequisite for a peaceful increase in human welfare on earth. The peoples of many DCs have already reached net reproduction rate 1.00 and within a few decades are likely to register zero population growth or a population decrease. Even in most LDCs the net reproduction rate is declining, in some countries quite rapidly, but it will take several decades before they reach net reproduction rate 1 and maybe a hundred years before the plateau is being reached. More and more often the future world population is being discussed in a model suggested by Frejka: when will net reproduction rate 1 be reached and when will the plateau (zero population growth) be reached?

Neither is there any reason to assume that the consumption of a mineral should follow an exponential curve except for, in the time scale of history, rather short periods. (See some of the many diagrams in this book.) It is for various reasons rational to assume that the period 1950–75 in the historic perspective will turn out to have been such a period of exceptionally rapid increase. The diagrams in the book indicate that pure S-curves are rather unusual. Curves that approach the S-form by way of platforms seem to be more common. The periods 1870–1910 and 1950–70 show a strong expansion while the period 1910–50 for many products were a period of stagnation. For the DCs most curves should reach a plateau within a few decades. It is not likely that man in the postindustrial society should devote almost all his active time to produce or consume material things.

Energy

Energy supply refers to man's external utilization of energy to heat, cool or light buildings, to drive machines and processes and so on, but not to man's internal utilization of energy to keep the body going nor to energy spent by nature. Plants directly utilize solar energy and animals, including man, do it indirectly by eating plants and each other. The 2 500 calories needed to keep a modern urban man going corresponds to 3 kWh per day (a 100 W lamp).

For a discussion of world consumption of energy Q is an appropriate measure. It is 10^{18} Btu or $2.52 \cdot 10^{20}$ cal or $293 \cdot 10^{12}$ kWh, which corresponds to 25 billion tons of oil. World consumption of energy in the beginning of the 1970s was ¼ Q per year and would, extrapolated at the prevailing growth rate (some 5 percent a year), be Q at the next turn of the century.

Tab. 4 – World Energy Consumption and Resources in the Year 2000

World consumption of energy in 1972:	$5.5 \cdot 10^9$ (oet)	
Estimated world consumption in 2000:	$25 \cdot 10^9$ (oet)	[1.0 t oil = 1.5 t coal]
Reserves	Oil equivalent tons (oet)	Times consumption in the year 2000 (Q)
Fossil fuels:		
Coal, oil, gas, shale	$5 \cdot 10^{12}$	200
Nuclear fuels:		
All fissionable fuels*, present reactors	$0.1 \cdot 10^{12}$	4
Same, breeder reactors	$15 \cdot 10^{12}$	600
Fusion, conversion of all deuterium in the sea	$35000 \cdot 10^{12}$	1 400 000
Other sources:		
Insolation	$125 \cdot 10^{12}$	
Water power	$25 \cdot 10^9$	'Income' per year, not
Geothermic energy	$0.1 \cdot 10^9$	'capital' as above
Tides, recoverable	$0.05 \cdot 10^9$	

* 10 million tons of U_3O_8.
Source: Geoffrey Chandler, 'Energy' in K. A. D. Inglis (ed.), Energy: From Surplus to Scarcity? (Barking: Applied Science Publishers, 1974).

Solar insolation over the globe is 5 200 Q per year of which 1 200 Q per year enter the hydrological cycle which is the solar radiation of prime concern to us on the surface of the earth. It has been suggested that if energy consumption reaches 1 percent of the latter quantity, say, some 10 Q per year, the climate of

16 Energy

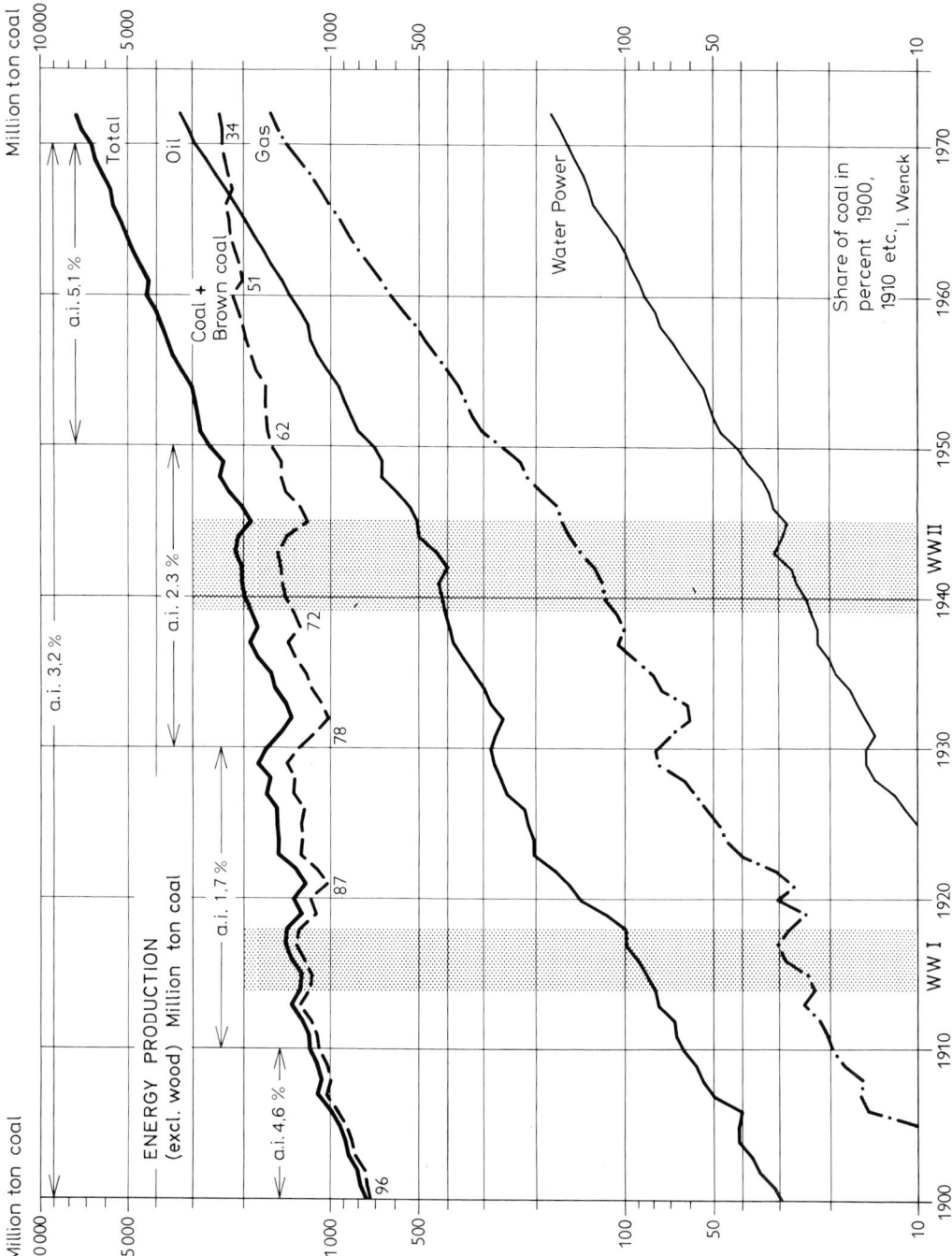

Fig. 5 – Energy production in the world 1900–1973 expressed in coal equivalents. Under the coal curve, the percentage of coal in the total energy balance is given for years ending with a zero. The years of the two world wars are indicated by a grey shading in the diagrams of this series.

the world would be affected. If this hypothesis proves to be true, this should provide the ceiling of energy consumption assumed by the logistic curve. In that case the energy curve must start to flatten out in the first half of next century.

We should be concerned about our oil resources. Fossil fuels will last into the 22nd century and possibly longer but will become gradually more expensive if they continue to play the role of a major fuel. The only alternative now available to the fossil fuels is uranium with thorium as a supporting fuel. Direct solar energy provides an alternative for heating buildings but is more doubtful as a source of concentrated high temperature energy.

In Sweden and some other countries accumulated solar energy in the form of wood was utilized during the Second World War in cars powered by producer-gas. But photosynthesis has an extremely low efficiency and the internal combustion engine has a low one. Before insolation has been converted into energy on the road in a car energy losses completely dominate. The utilization of wood as a fiber material and a construction material provide such high alternative costs that wood can be used as a motor fuel only in an emergency. The handling of wood is so laborious (and wages so high) that even the farmer in the middle of

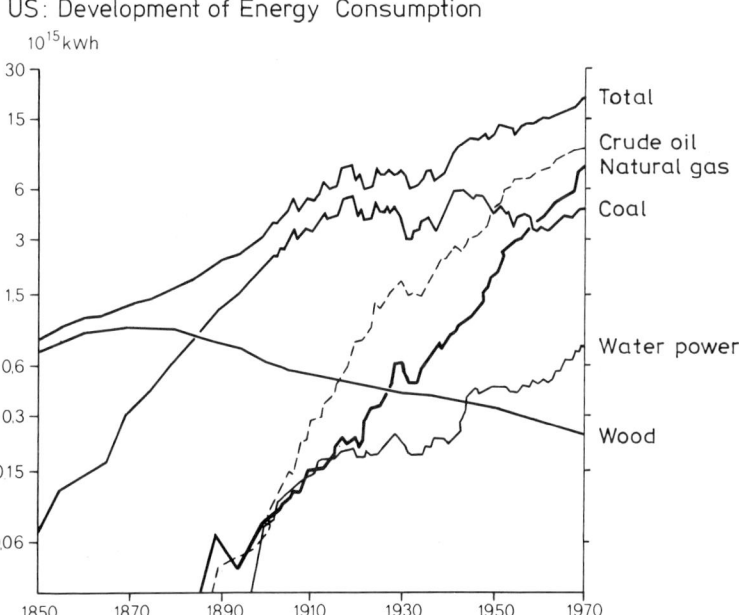

Fig. 6 – In the American pioneer society wood played a prominent part as a source of energy. Wood and charcoal are still important in many LDCs. River steamers and locomotives were fired with wood before coal became widely accessible and wood was a common household fuel. In the modern society wood is too valuable as an industrial raw material to be burnt; in addition it is expensive to handle. Similar arguments will soon be heard about the use of fossil fuels as a source of energy when nuclear power is available.

18 Energy

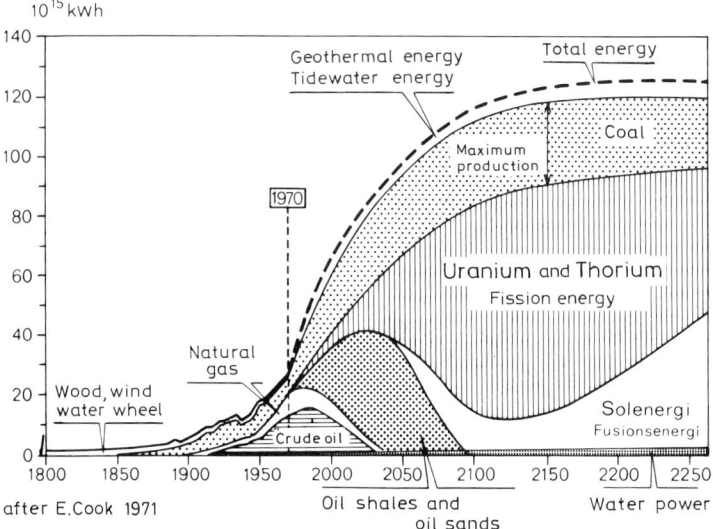

Fig. 7 – The sketch shows the American energy consumption being adjusted along a logistic curve to a per-caplita-level about four times the 1970 consumption. The relations are realistic, seen with the knowledge and expectations about technical breakthroughs soon to come, prevailing around 1970. With a somewhat higher standard of living than Sweden, the United States about 1970 had twice as high energy consumption per capita. With higher energy prices, the American plateau is likely to find a considerably lower level than the one shown in the sketch. Investments in energy saving techniques became interesting after the 1973/74 oil crisis.

the forest, with free access to wood, has an electric oven in his kitchen and an oil furnace in his basement.

Production of primary energy in the period 1910–30 had an annual growth rate of 1.7 percent, figure 5. This rate increased radically to 5.1 percent in the 1950–70 period. The rate was slightly higher in the 1950s than in the 1960s (5.1 and 5.0 percent, respectively). Coal saw its share of total primary energy, except wood, decline from 96 percent at the turn of the century to 34 percent in 1970. Some primary energy production can be seen as consumption or input at the next step. As fuel efficiency gradually has been improved since the beginning of the Industrial Revolution, energy production at the following step (output) has an even higher rate of increase.[3]

Fossil Fuels

Fossil fuels, found in sediments that are dated with the help of fossils, are also known as mineral fuels, since they are obtained in the mineral realm although they have their origin in the 'vegetable' or 'animal kingdoms'. They are coal, petroleum and natural gas.

Tab. 5 – Estimated World Resources of Fossil Energy

Oil	7–11 Q
Natural gas	10 Q
Oil sands and shales	30 Q
Coal	200 Q

Coal

Coal is a solid, black mineral made up of carbon, hydrogen, oxygen and nitrogen in varying proportions. In addition it contains impurities such as ash and sulfur. From early in the Industrial Revolution coal was a major fuel competing with charcoal and wood. The steam engine gradually superseded direct water power as a source of manufacturing energy. Coal initiated the high exponential growth of energy consumption per capita that still goes on.

Coal fields and receiving ports for coal strongly influenced the location of manufacturing industry and population in industrial countries. Cost of land transports was long very high. As seen from figure 5 the role of coal as a source of energy has been drastically reduced with the expansion of oil and natural gas in this century. Since the reserves of coal are much larger than those of petroleum, coal may again be expected to play a larger role in the future. Coal is a substitute for oil and natural gas in many uses and it may be used for making synthetic gas, gasoline and other hydrocarbons. A radical shift from oil to coal at the electric power plants in the world would by 1985 save 250 million tons of oil per year and it would cost 375 million tons of coal.

Origin of coal – Coal is of vegetable origin, mainly treelike ferns, rushes, and lycopods that grew in swamps and lagoons and went through a peat stage. Over 3 000 plant species have been identified from Carboniferous coal beds. The lush vegetation and the large growth (ferns and lycopods attained some 30 m in height) indicate a mild temperate to subtropical climate with moderate to heavy rainfall distributed throughout the year similar to that of the Carolinas in the present United States. Coals have been found in Spitzbergen, Greenland, and Antarctica where they would not now be formed. Coal seams are interstratified with shales, clays, and sandstones and are usually underlaid by an underclay containing roots of plants.

The change from plant debris to coal involves biochemical action, preservation of the material from further decay, and pressure under accumulated plant materials and other later sediments. Coals are classified according to the stage of transformation from vegetation to fixed carbon. In the northern Hemisphere coal seams are usually Carboniferous and in the southern Hemisphere Permian in age, but coal occurs in all post-Devonian periods. The Tertiary yields most of the lignite of the world.

Types of coal — Carbon is present as solid (fixed) carbon and in volatile matter (gas). In addition coals contain varying degrees of moisture. From the highest rank downward, in terms of fixed carbon, coals are classified into anthracite, semi-anthracite, super bituminous, bituminous, subbituminous, brown coal and lignite. Anthracite and bituminous coal are referred to as hard coal and brown coal and lignite as soft coal.

Diamond is pure, naturally crystallized carbon of extreme hardness and graphite is soft, pure carbon. Both forms are purer than anthracite. It is only the form of the crystal structure that separates diamond from graphite.

Peat has lower carbon content than lignite. It is an accumulation of partly decomposed vegetable matter and represents the first stage in the formation of all coals. It is used as a fuel in some parts of the world, even as a fuel in power plants (Ireland, USSR), but its main use is as a soil conditioner.

Lignite represents the second stage. Sometimes a distinction is made between lignite and brown coal, the latter being unconsolidated and the former consolidated. Lignite is relatively young and contains much moisture but little fixed carbon (some 38%) and volatile matter (19%). Its heating value is only half that of high-rank bituminous coal.

Bituminous coals of various types are low in moisture (3–23%), and high in fixed carbon (42–83%), volatile matter (12–41%) and heating value. They are the most used and desired coals. They are classified as noncoking coals (12–20% volatile matter), coking coals (20–30%) and gas coals (over 30% volatile matter). Gas coal is used for the manufacture of city gas, with coke as a by-product. Coking coal is used in coke ovens for the manufacture of metallurgical coke, with gas as a by-product.[4]

Anthracite is a jet black, hard coal with high luster. It represents the highest grade and the most metamorphized coal, over 95% of carbon, formed under the greatest stress. It is used almost exclusively for domestic heating.

Cannel coal falls outside the mentioned series. It was probably formed in ponds; it occurs in lenticular masses and often contains fossil fish in addition to windborne spores and pollen. This variety of bituminous coal burns with a long flame and is preferred for fireplace coal.

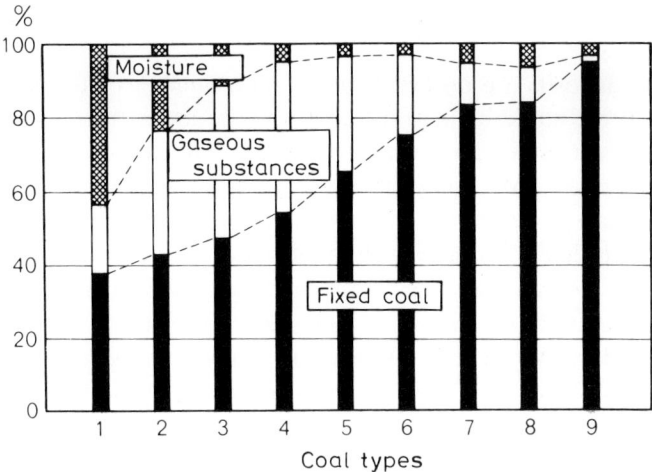
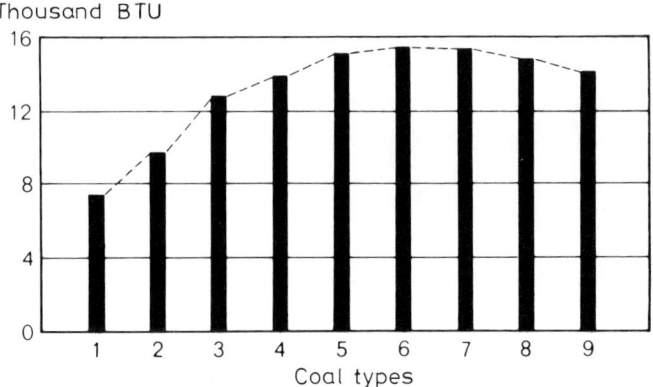

Fig. 8 – The structure of the coal types (top) and their heat value (bottom) expressed in British Thermal Units (BTU).
1 – Lignite and brown coal were formed in late geologic times and contain much water. Their heat value is only half that of some coals. 2 – Subbituminous coal. 3–5 – Bituminous coals, which have their name from their high content of gaseous hydrocarbons and tar (bitumen). 6–7 – Semibituminous coals. 8 – Semianthracite and 9 – Anthracite.

Coal is further classed into grades according to the amount of impurities, such as rock, sulfur and ash. Little coke is made from coals having over 1.5% sulfur. Coal is often upgraded by washing.

Occurance of coal – The thickness of coal seams ranges from a mere film up to 30 m. Some seams are remarkably flat and persistent. The Pittsburgh seam, for instance, underlies 100 000 km². Most seams vary considerably in thickness. Generally, a given vertical section contains several coal seams, some times as many as fifty or more, many of which may not be workable. Anthracite coals are found in regions of close folding and bituminous coals and lignite in regions of gentle or no folding.

Whether or not coal deposits are utilized depends on a number of factors. Location with regard to the market is a prime concern. The lower the cost of transportation, the higher will be the revenue per ton at pithead. Depth of working area and thickness of coal seams are also important. More and more coal is being mined in open pits. Examples are the huge coal mines in Queensland at which the overburden is scooped with giant power shovels taking as much as 200 tons per bite. Frontloaders then scoop up the coal into trucks to be carried to a loading terminal where it is dumped into huge railroad cars. Heavy unit trains run the coal down to the coast where it is dumped into giant bulk carriers for transport to Japan or Europe. European mines as a rule are underground and often have thin seams, some of which only allow pick and shovel mining.

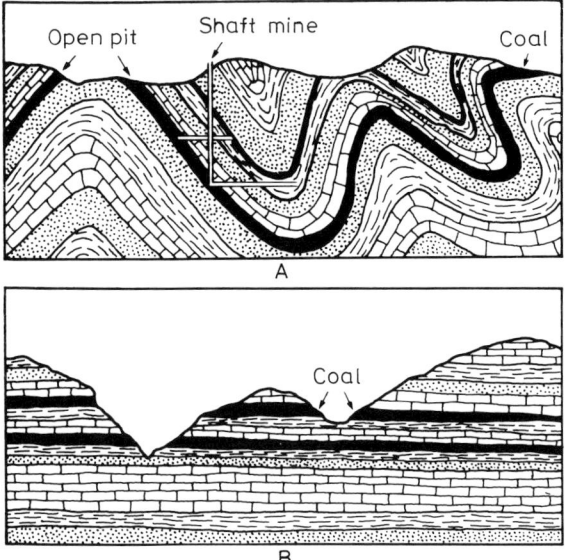

Fig. 9 – Profiles through folded strata with anthracite seams (A) and through horizontal strata with bituminous seams (B). Mining is more expensive in seams of the first type, found e. g. in northeastern Pennsylvania. The horizontal seams of the Appalachian Plateau can often be mined by way of a tunnel (drift, slope, incline), in open-pit mines which allow heavy mechanization, or, in sloping land where the overburden is too thick, in auger mines.

Overseas coal from the United States, and more recently from Australia have been in a position to compete with domestic coal in several European countries. The former European export industry has survived on a reduced scale thanks to state subsidies voted through parliaments on social grounds. In OECD-Europe the share of coal in primary energy consumption was reduced from 60 percent in 1960 to 29 percent in 1970[5].

The type of coal is also important. Most coals have competitors but coke is indispensable for the blast furnace and good coking coals are in short supply. Eventually, rising coke prices will force the steel industry to bypass the blast furnace but direct reduction methods are still uneconomic in use.

Only 7% of the world's coal production enters international trade, most of this coking coal from North America and Australia to western Europe and Japan. This is in great contrast to the beginning of the century, when coal ruled supreme as a source of energy and was a major commodity in international trade.

Lignite, which varies a great deal in quality, can be transported only in dehydrated and compressed form, as lignite briquettes. Most lignite is used as fuel in thermo-electric stations.

Europe

United Kingdom – Large coal deposits and an early start were the two pillars on which Britain built its position as the workshop of the world in the early stage of the Industrial Revolution. Th. Newcomen in 1712 constructed a practical steam engine based on principles presented in France in the 1690s. It was built for lifting water from mines. James Watt after 1769 improved the mechanics of the steam engine so that rotating movements became possible. The efficiency of the steam engine was improved through a series of inventions, first in Britain and later also in other countries. The Swiss Ch. Brown in the 1860s built steam engines with a coal consumption of 2 kg/kWh as against 20 kg for Newcomen's machine. The steam engine, now a museum item, represents a breaking point in man's development, to be compared with the introduction of agriculture in the beginning of the neolithic period, some 10 000 years ago. With its successors, the steam turbine, the combustion engine and the gas turbine, the steam engine opened the way for a continuously increased energy consumption per capita, a development that was unimaginable for preindustrial man. A relative stagnation was turned into exponential growth.

Coal was the large energy base not only for the heavy industries, such as iron and steel, but for light industries as well, for instance the textile industries. The undeveloped land transport system of the day made a location close to cheap coal imperative. Coalfields and harbors became the natural locations for light and heavy industry. As late as 1860 Britain accounted for more than half the world production and the British output continued to increase until 1913, when it reached an all-time peak of 287 million tons of which 100 million tons were exported. In 1860 the United States produced less than one fifth as much coal as Britain but the American expansion rate was high. At the turn of the century

24   Energy

United States exceeded Britain as a coal producer. In the 1930s, the British coal mines employed almost a million people and accounted for 90% of the British mineral production by value. About 1970 these figures were 300 000 and 80%. The decline in employment has caused serious adjustment problems in the coal districts.

There are six major coal field areas in Britain: the Scottish Lowlands (1), the Northeast area or Northhumberland-Durhamshire (2), the east (3) and west (4) flanks of the Pennines, the Midlands (5) and South Wales (6). (Figures refer to orientation map, figure 10)

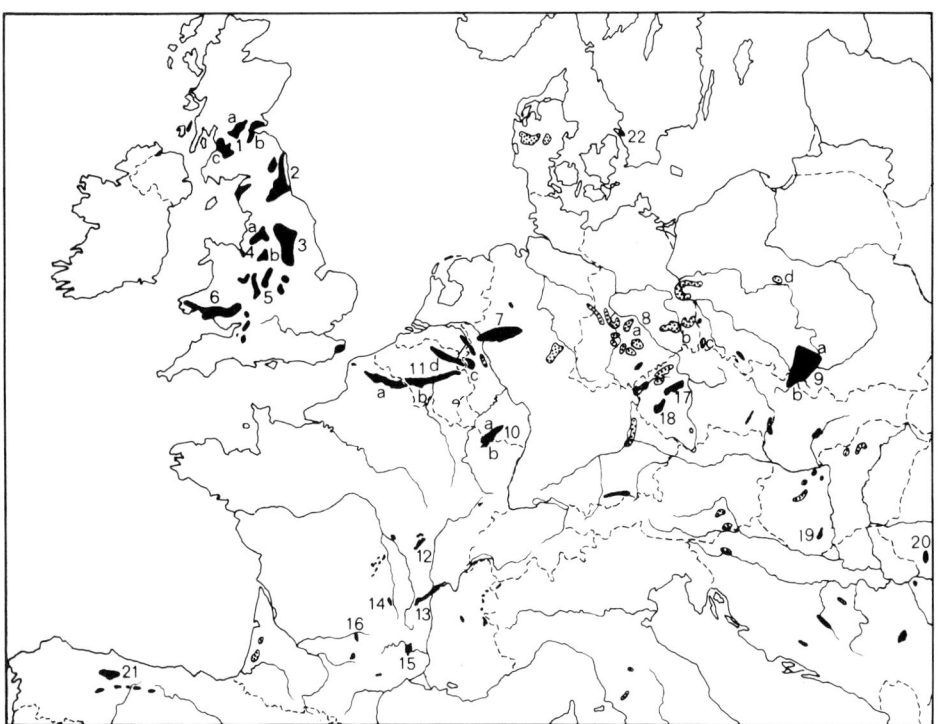

Fig. 10 – Orientation map of coal fields in Western and Central Europe. Figures and letters refer to the text.

The Scottish Lowlands contain three fields. The area southeast of Glasgow centered on Lanark produces coking coal (1 a). This was the first center of the Industrial Revolution in Scotland. Near Edinburgh a field stretches under the Firth of Forth and good steam coal is mined on both sides (1 b). The third field is on the west coast, in Ayrshire near Irvine and Ardrossan (1 c).

The coal fields on the northeast coast of England produce both steam and coking coal and have been worked since Medieval times. They are favorably located for shipment up and down the coast and overseas. The hauling of coal from this area to London was for centuries one of the busiest coastal trades anywhere. The first railroad, between Stockton and Darlington, was built in 1825 to haul coal from inland pits to the coast. The Northeast was a major participant in the British coal exports which reached a peak in 1923 with 86 million tons. "Carrying coals to Newcastle" was long the ultimate in doing the unnecessary. After the Second World War West European coal fields, with their thin, faulted, and − in France, Belgium, and the Ruhr − highly inclined seams, have found it increasingly difficult to compete with American coals, produced under more favorable conditions. Imported American coal has actually been unloaded at Newcastle in the postwar years. Productivity in the British mines increased in the 1960s from 1.4 tons per manshift to 2.2 tons but the American output in 1970 was 18 tons per man and shift.

The Yorkshire-Derbyshire-field on the east flank of the Pennines is the leading field in the country, accounting for increasing shares of the national production, almost 50% of the 1970 output of 145 million tons. The old, small and uneconomic mines, which have rapidly been closed down, are located in the western part of the area and new and large ones in the east, e. g. around Doncaster. Three manufacturing regions receive coal from this field. The mines in West Riding deliver coal to the wool industry at Bradford, Halifax, Huddersfield and other towns. Southern Yorkshire supplies the iron and steel industry at Sheffield and Doncaster, and Nottinghamshire-Derbyshire has been the energy base of the diversified industry of Nottingham and Derby.

The fields on the west flank of the Pennines are deeper and more faulted which makes for higher cost in mining. These fields were of great importance for the large cotton industry of Lancashire, which had its peak before the First World War and is now only a small fraction of its former self. In North Staffordshire the coal fields have been of great importance for the famous potteries.

In the Midlands, centered on Birmingham, coal seams are thick but deep. Mines are operating as deep as 1000 m. Much of London's coal supply comes from this area, which early became known as the Black Country.

The South Wales field, exploited since the 13th century, is the only British coal district with large amounts of anthracite, located in the western part of the area. The coal field reaches the coast at Swansea but is a few miles inland at Cardiff and Newport. Mines were built in the deeply incised river valleys and the coal hauled by rail down to the coal piers on the coast. The mines and the heavy industry, oriented to the coal, form one of the major British industrial districts, but this is not much in evidence from the bog-covered high plateau where sheep herds graze. That landscape is also typical of South Wales.

Newport in England is the oldest coal port of the area. It was connected with the coal field by canal about 1800. A few years later Swansea started to ship coal. Cardiff and its neighbour ports, Penarth and Barry, get their coal by rail. They dominate modern coal shipments from South Wales. When ships were propelled by coal-fired steam engines Cardiff and Newcastle were known as major suppliers of bunker coal to ports and bunker stations around the world.

Continental Western Europe – Continental Europe is richly endowed with coal and with rivers and canals that provide low-cost transport. The major fields all lie north of the Alps. Mediterranean Europe and the Nordic countries are poor in coal. The five major coal fields are the Ruhr district (7), the lignite fields of central Germany (DDR, 8), Upper Silesia (Górny Śląsk, 9), Saar (10), and the Sambre-Meuse Valley (11), connecting the coal district of northern France (11a) with that of Aachen (11c) in Germany. As in Britain, the Continental coals fields and their transport systems have made a great impact on the population map. Industrial settlements cluster on the coal fields and along the rivers and canals. The vast conurbation in the Ruhr district compares with London and Paris in size. The geological formations that form the base of vast industrial complexes run across national boundaries in several instances. The postwar attempts at unification of western Europe according to Jean Monnet's vision – to avoid future wars in the densely populated heartland of the Continent – started with the important coal and steel complex. The European Coal and Steel Community (ECSC) was established in 1952.

Near the Ruhr, a small tributary of the Rhine, coal has been mined on a small scale since medieval times from seams exposed by erosion. From about 1850 heavy industry was attracted to Europe's largest coal field. Coal production increased from 0.2 to 1.7 million tons between 1800 and 1850 and grew to 55.1 million tons in 1899. Germany (like France and the United States) was not so hard pressed as Britain to switch to coke-fired blast furnaces, which produced a cheaper but less versatile iron than the charcoal furnaces, which until the second half of the 19th century were scattered over the forested parts of Europe. The construction of railroads created a demand for mass-produced steel; the new steel processes patented between 1856 and 1878 made it possible to mass-produce that steel.

The coal seams dip relatively steeply northwards from the Ruhr River, where the exposed coal is lean or anthracitic. More valuable coal with a higher volatile content can be reached only by shaft mines which become progressively deeper northwards in the Hellweg, Emscher, and Lippe zones. The average depth of mines in the Ruhr district is now almost 800 m. In 1974 only 88000 miners were employed in underground operations.

About one-fourth of the coal is used by industry within the Ruhr district. Much is shipped from Duisburg-Ruhrort to coal ports along the Rhine. The Rhine Valley, including the Ruhr, is the manufacturing heartland of Continental Europe and the Rhine River one of busiest waterways in the world. As a source of energy and as a raw material for the chemical industry, coal meets serious competition from oil and natural gas also in the old Coal Valley, which once dominated the chemical industry of the world. But there is no substitute for coke in the blast furnaces.[6]

The Saar field is being mined both in Germany (10a) and France (Moselle, 10b). The coal, some 10–15 million tons a year in each country, is being used by the heavy industry of the Saar and of Lorraine. The main French coal field in the departments Pas de Calais and Nord produces some 20–25 million tons a year or about half the French output. France has some small coal fields in scattered locations, some of very old standing, such as the field at Le Creusot (12) and St. Etienne (13). Other fields are at Clermont-Ferrand (14), Alès (15), and the Tarn River (16) northeast of Toulouse.

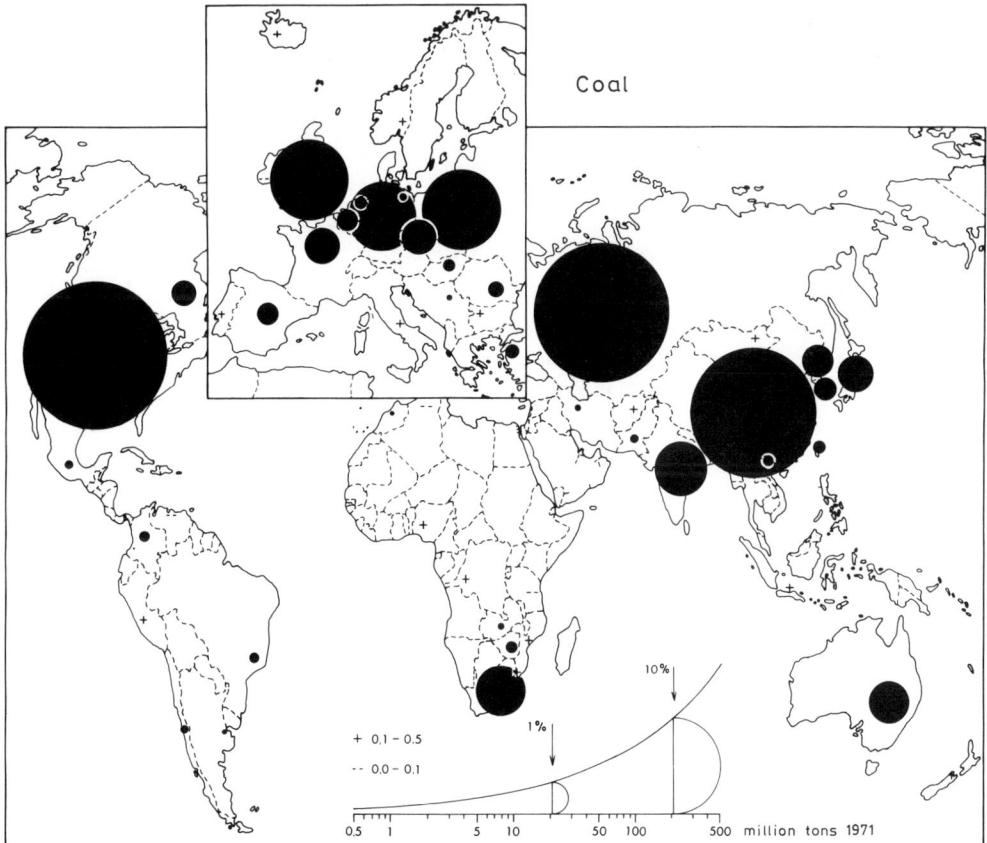

Fig. 11 – The dominating position of Europe as a coal producing region has been lost with the expansion of output in regions more favored by nature: United States, Soviet Union, China, Australia and South Africa.

The field of northern France is an extension of the Ruhr coal measures forming a ten-mile-wide arc from Cologne to Liege. In Belgium it runs along the Sambre and the Meuse, where the old Belgian heavy industry is located, centered at Mons−Charleroi−Namur (the Borinage field) and at Liége−Seraing. In the interwar years a new and deep field was opened in northern Belgium, the Campine field, which continues with the Dutch field at Maastricht in the Province of Limburg and the old German field at Aachen. For various reasons productivity is low in the Sambre-Meuse-Region and the Wallonian share of Belgium's coal output is on the decline. In 1968 Campine accounted for 57% and Wallonia for 43%. Coal production in Limburg in the late 1960s was more than half the Dutch consumption, but these mines were expected to be closed down about 1975 because of competition from cheap natural gas.

Central Europe − Among the countries of Central Europe, Poland is the major coal producer. It had a large share of the fields in Sląsk (Silesia) in the interwar years. After 1945 the former German part of the coal fields came under Polish control. The heavy industry at Gleiwitz−Beuthen−Katowice was established already in the 19th century. Coal production in this field was sufficient also for exports. Gdynia was developed as a Polish coal port from about 1925. In the 1960s and early 70s, Poland was a major coal exporter, on a par with West Germany and the Soviet Union and surpassed only by the United States. Shipments to Comecon countries primarily went by rail.

Czechoslovakia has also a share in the Silesian coal fields with mining concentrated primarily around the industrial city of Ostrava. Minor Czech coal fields are located at Kladno and Plzen (Pilsen). East Germany has some coal production in Saxony, Hungary at Pécs and Romania in the southwestern Province of Hunedoara.

The vast deposits of brown coal at the foot of the central German highlands account for almost half the world production of this mineral. In 1970 some 261 million tons were produced in East Germany, 108 million tons in West Germany, and 33 million tons in Poland. The large dead weight of ash and water makes it impossible for brown coal to be transported from the field. The coal is mined with giant excavators and is fed directly into large thermo-electric power

→

Fig. 12 − The curves of coal production 1860−1973 were based on several sources: first Sundbärg, who includes brown coal (1), followed by Weltmontanstatistik through 1920, and, from 1921, the League of Nations and the United Nations. The break at 3 shows the world, excluding China, and at 4 the world, excluding China and the Soviet Union. Germany with its large brown coal production has a conspicuous break at the shift from Sundbärg (1).
Germany in 1870 produced 7.6 million tons of brown coal, in 1880 some 12 million tons, in 1890 some 19 million tons and in 1900 some 40 million tons. At 2, German production is shown for the prewar territory and thereafter for the interwar area. France at 5 shows the country excluding Lorraine.

Fossil Fuels 29

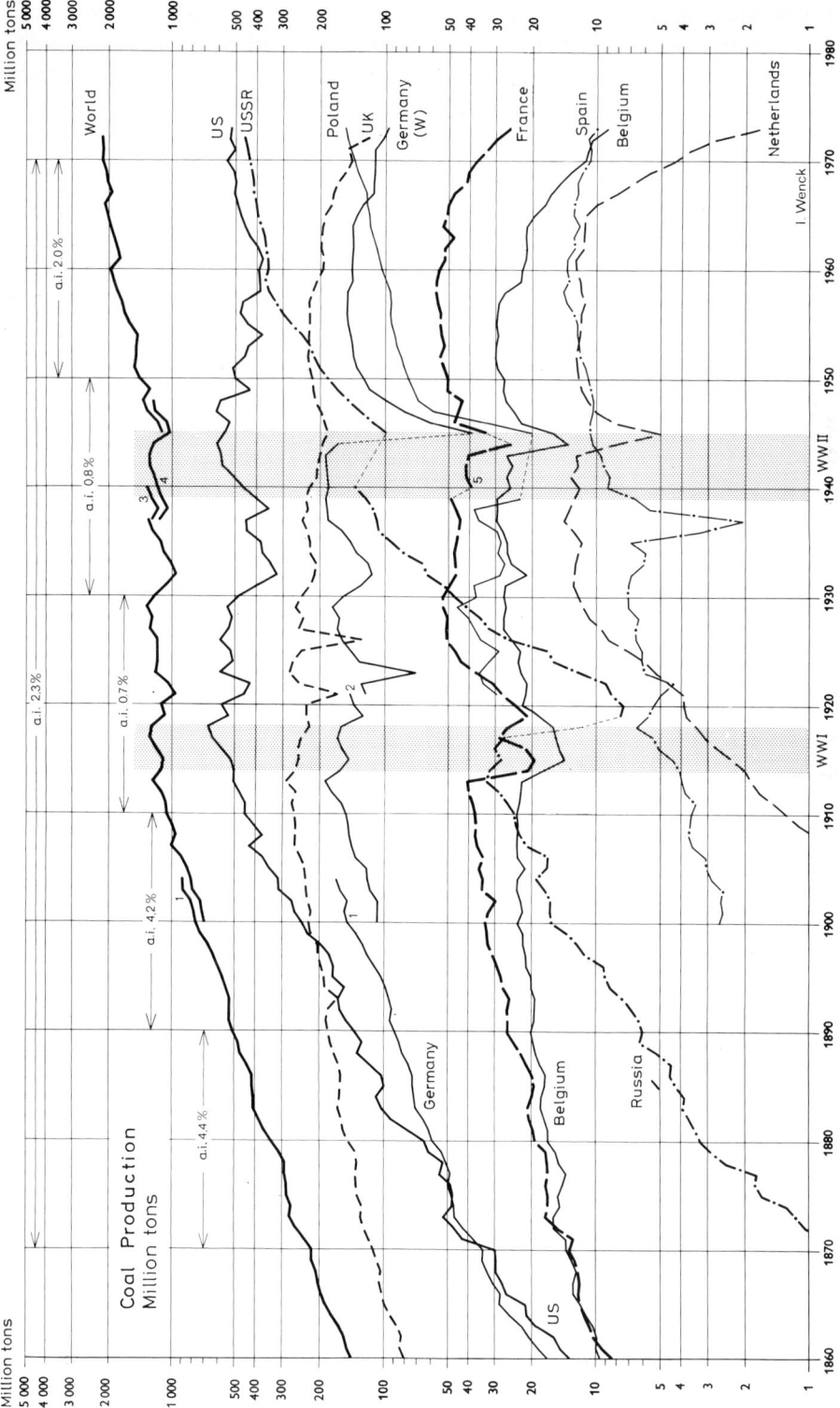

plants with turbines according to the counter-pressure principle, in which the steam is not condensed but is made to move to an adjacent briquette-works, where it is used to dehydrate brown coal to briquettes. In the interwar period energy intensive chemical complexes were located at the brown coal fields, e g for making synthetic gasoline, synthetic rubber, fertilizers and aluminium.

Brown coal briquettes are a dominating household fuel in East Germany. Brown coal accounts for almost 90% of the area's electricity. Brown coal, water power, and nuclear plants take the base load (on the average 7000 hours a year for a brown coal plant in West Germany), coal plants take the average load (1500–5000 hours a year), and oil and gas plants take the top load (less than 1500 hours a year).

The vast brown coal fields of West Germany are between Köln and Aachen, not far from the small coal field at Aachen, and those of East Germany are located in the Halle-Leipzig-district (8a) and east of the Elbe River in Niederlausitz (8b). The Polish fields are at Turoszow (8c) and Konin (8d).

Other Europe – Spain is the major coal producer among the Mediterranean countries. The main field is located in the north at Oviedo in Asturias. Spain is almost selfsufficient in coal but productivity is low and the coalfields are known as an area of social unrest. Among the Nordic countries Norway has some coal production in Spitzbergen, where also the Russians through their organization Arctic Ugol' have the right to mine coal and prospect for petroleum. The small scale mining at Höganäs (Sweden) has been discontinued.

North America

The largest coal reserves in North America are located west of the Mississippi, on the Great Plains and in the intermontane basins of the West, but production has been dominated by the East, primarily the Appalachian region. As in Europe, coal was the major workhorse in economic life through the Second World War and some of the postwar years.

United States – Oil exceeded anthracite in 1918 and coal in 1952 expressed in amount of energy. Natural gas surpassed coal in 1958. The former energy giant now holds a poor third position in the United States.

The mining of anthracite is almost completely confined to a small area of northeastern Pennsylvania in the Ridge and Valley Country around Scranton and Wilkes-Barre. Anthracite production first passed the million-mark in the 1840s and exceeded bituminous coal production until about 1870, when some 20 million tons were produced of both types. Anthracite reached its peak in 1917 with 100 million tons, while bituminous coal peaked at almost 600 million tons in the following year, a level not attained until World War II and the immediate postwar years.

The smokeless anthracite has long been a popular home fuel but in the 1920s it ran into competition from fuel oil and later from natural gas. In the 19th century it was also used as a fuel in the iron industry which until 1875 had its center near the anthracite region. Since anthracite mining is geographically concentrated and very labor intensive, in the 1920s employing more than 150 000 people, it became the base of the strong miners' union.[7]

But the anthracite industry faced an adverse structural change; production dropped to less than 10 million tons and employment to some 7 000 in 1970. The mining region, which early attracted silk and apparel industries from the New York area, became one of the outstanding depressed districts in the United States. However, through combined efforts of local, state and federal interests it was possible to reactivate the region from about 1965. The new system of national motorways made the area easily accessible from greater New York; the grain of the country and the coal railroads point to Philadelphia which is further away.

Bituminous coal is produced primarily in the Appalachian coal measures extending west of the Allegheny Mountains from western Pennsylvania to central Alabama. By comparison with Europe, conditions for mining are exceptionally favorable. The seams are almost horizontal, they occur at moderate depths, they are often accessible from the sides of deeply incised river valleys, and they are usually 1.5−3 m thick, which early allowed underground mining and loading to be mechanized. The output per man-day which was 4 tons in 1920 and 6 tons in 1946 started to climb steeply about 1950 and reached 18 tons in 1970.[8] The ever-larger dragline cranes made it feasible to remove the overburden and scoop up coal in strip mines. Open-pit mining accounted for 24% of the US output in 1950 and for 40% in 1970.

The American bituminous coal industry has seen most of its traditional markets disappear after the Second World War. The railroads alone consumed as much as 100 million tons during the peak years of the 1940s; now this market is nil. Diesel electric engines were more efficient than steam locomotives which meant a switch from coal to oil. For many uses fuel oil and natural gas are superior to coal which became obvious in the United States already in the 1920s. However, in two markets coal is highly competitive: as coking coal and as a fuel in large stationary units such as thermo-electric power plants and cement factories.

Until the blast furnace can be bypassed in steelmaking, coke is needed by the iron and steel industry. Coking coal is predicted to be in short supply in the world within a few decades. The direct reduction method is one way of solving this problem, to mix non-coking coal with coking coal is another. Research along both lines is carried on in several places of the world. A direct reduction method was sketched already by Sir Henry Bessemer but the practical problems of making the method profitable have turned out to be very difficult.

The most important remaining market for coal is as fuel in thermo-electric power plants, in which coal is competing with fuel oil, natural gas and lignite. While other markets have dwindled, this one has expanded, which explains why the American coal production declined from a peak of more than 600 million tons around the end of the Second World War to a low of 379 million tons in 1961, when it barely exceeded the Soviet output, and then rapidly increased again. Since the two remaining markets, and especially the production of electricity, have a high growth rate, American coal production already before the oil crisis of 1973/74 could be expected to reach new record levels in the near future.

Legislation against air pollution (The Clean Air Act, 1970) made many stationary units that used coals with a high sulfur content switch to fuel oil low in sulfur, imported from Libya and other Arab countries. But this might have been a temporary change even without the oil crisis. A new type of power stations, combining gas and steam turbines, promise to be more efficient than conventional stations. Gasified coal or power gas, which has a low heating value compared with natural gas or synthetic gas (only $1/_7$) and therefore must be used at the site of production, promises to be the cheapest way of eliminating sulfur from coal combustion. The first power station of this type was started at Lünen, Germany, in the early 1970s. Several American orders for the German Lurgi-system were placed during the oil crisis and several American systems were being tested.

The Appalachian coal deposits, covering an area of some 175 000 km^2 and stretching from northern Pennsylvania to central Alabama, has yielded more coal than any other contiguous coal formation in the world. The natural

Tab. 6 – International Trade in Coal, Coke and Briquettes 1970 (SITC 321) (million US $)

Imports		Exports	
Japan	1 015	United States	1 044
France	352	Germany (W)	682
Belgium	302	Poland	408
Italy	211	USSR	407
Germany (W)	177	Australia	210
Canada	157	Netherlands	81
USSR	124	Great Britain	70
Netherlands	123	France	63
World*	3 217	World*	3 141

* excluding Socialist Countries
Source: United Nations, Yearbook of International Trade Statistics 1972–73

conditions for coal mining are very favorable in the dissected Appalachian Plateau, where horisontal coal seams are found at shallow depths, often exposed in the many river valleys, allowing for the low-cost drift type of entries. Instead of a vertical shaft, commonly used in Europe, a tunnel (drift or slope) is built to the mining area. Underground the coal is usually extracted by the room-and-pillar system, which is labor-saving but coal wasting. About one-third of the coal is left in the form of pillars. As seams are thick, both cutting and loading were early mechanized. Since about 1950, increasing amounts of coal have been mined in strip mines which has led to drastically increased outputs per man-day but also to large ecological problems.

In the Appalachian field there has been a shift of the center of coal production from the northern area – south-western Pennsylvania and adjacent parts of West Virginia and Ohio with the Monongahela and Ohio Rivers as the major coal carrying waterways – to an area straddling the borders of West Virginia, Kentucky, and Virginia. Since the turn of the century this southernly area has been connected with Toledo and other Great Lakes cities and with Hampton Roads by railroads specially designed to carry coal. On a map of railroad cargo in the United States they stand out with their heavy traffic. The Kanawha River is a major waterway in this district. The Appalachian field contains good coking coal in the eastern part, e. g. the Connelsville coal of western Pennsylvania and the Pocahontas coal of southern West Virginia. An outlying mining district within the Appalachian formation is centered in Birmingham, Alabama. The mining population of the bituminous coal districts have traditionally lived in small towns. Very few exceed 10 000 people in size. From a social point of view, coal mining has long been a problem industry in spite of expanding production. In the United States employment fell from 415 000 in 1950 to 140 000 in 1970. The coal mining districts are economically depressed areas.

Favorable location with regard to the densely populated Manufacturing Belt makes the Eastern Interior Field the second most important. Coal is mined chiefly in southern Illinois and Indiana and western Kentucky. Seams are near the surface and most of the production has long been from open pits. Unit trains run in shuttle traffic between the loading terminals and thermo-electric power plants on the outskirts of the big cities.

The Western Interior Field stretches through Iowa, Missouri, Kansas, Oklahoma and central Texas. Coal beds are near the surface but coals are generally of lower quality than further east. The fields are further away from the large northeastern markets and face strong local competition from oil and natural gas. Good coking coals in eastern Oklahoma are used in regional steel plants (Texas) and some is shipped to Utah and California. The large lignite field stretching from Little Rock to San Antonio is exploited on a very limited scale.

The extensive coal formations in the Northern Great Plains, in the Rocky Mountains and near the Pacific Coast in the Northwest are mostly of low rank, to a large extent lignite. Distances from the market and in some instances rugged terrain have prevented development. Coal of the Book Cliff's field of eastern Utah and western Colorado, especially Sunnyside south of Salt Lake, is obtained in shaft mines to supply blast furnaces in the West. This coal is blended with better coking coal from Oklahoma to improve quality. Some production for local use occurs in several places in the West as well as in Alaska, where extensive coal formations are known to exist. But coal plays a very minor role in the energy balance of the states west of the Mississippi-Missouri, often just a few percent, while in most eastern states coal is still the leading fuel, ahead of natural gas and oil that dominates in the western parts.

Federal energy planners suggest a doubling of the American coal production to 1 200 million tons by 1986. Most of the increase should be west of the Mississippi. Two counties alone, Gilette and Campbell on the Powder River at the border between Wyoming and Montana, are expected to produce 100 million tons a year. Much coal should be converted to synthetic gas at the mine. The 'syngas' should be substitute for natural gas which is expected to run out by 1985. The rest of the coal should be converted to electricity at huge power plants on the coal fields.

The Sierra Club and other environmental groups have taken the issue to the courts. They want broad studies on the socio-economic impact of the largest investments ever made in the United States in a remote, sparsely populated countryside.

Many coal fields are owned by oil companies. Open-pit mining and large 'syngas' plants and thermo-electric power plants are expected to make this energy production profitable.

Canada – Some American coalfields continue on the Canadian side of the border. In the West, in areas underlaid by coal or lignite, many families used to dig coal for their own consumption as was done in many parts of the North American Continent. Local production on a larger scale also occurred. But the vast deposits of coal and lignite in the West meet serious competition from oil and natural gas and total production has remained small. However, recently large deposits of good coking coal near the Alberta-British Columbia border in the foothills of the Rocky Mountains have been developed for export to Japan. Over 10 million tons a year are shipped by way of the new Roberts Bank Superport south of Vancouver, opened in 1970.

Until about 1920 Canada's Appalachian deposits, mined in northern Nova Scotia and Cape Breton, supported the leading mineral activity in the nation. Since then, deposits of many other minerals have been developed and coal now ranks low by value. Production in the Sydney region, part of which is under the ocean, is subsidized by the government. Normally Canada imports over half its consumption of coal.

USSR

The Soviet Union is considered to have the world's largest coal reserves. At the present rate of consumption they will last 20 000 years. But $^3/_4$ of these reserves are in eastern and central Siberia and the Far East. Development has taken place as near as possible to markets.

The Russian coal production, 28 million tons in 1913, expanded very rapidly in the Soviet period from 30 million tons 1928 to 311 million tons 1955 and 451 million tons 1970. Underestimation of oil and natural gas reserves led to emphasis on coal in official policy until the 15-year plan for fuels of 1957 reversed this policy. In 1961 Soviet coal production almost equalled that of the United States but it has since grown at a slower rate. In 1950 coal accounted for 73% of fuel production but in 1967 for only 42%. Per heat unit the cost of oil production in the Soviet Union is only one fourth that of coal. Giving top priority to coal so long in the Soviet energy policy may turn out to have been a planning mistake as expensive as the West European laissez-faire-policy for oil which led to a strong dependance on Arab oil and the payment of a high monopoly tax to the OPEC countries after the oil crisis 1973/74.[9]

The increased oil production has relieved the Soviet railroads. While coal has to be shipped by rail, oil is partly distributed by pipeline. The rapid switch to electric and Diesel locomotives (more than 85% in 1970 as compared with 26% in 1958) has relieved the trains of carrying their own fuel. The railroads until 1957 accounted for more than 20% of the national coal consumption. As in the market-economy-countries thermo-electric power stations, cement factories and the steel industry are the remaining heavy users of coal.

Donbass – The leading coal field in the Soviet Union is still the Donets Basin, the Donbass, originally developed in the 1880s with French and Belgian capital. This coal field and the iron ore deposits at Krivoi Rog formed the basis of Czarist Russia's modern steel industry. It also played a prominent role in the Soviet 5-year plans but its relative position declined with the development of new coalfields. In 1972 it produced 32 percent of the country's coal and more than half its coking coal. Some 95 percent of the output in the Donbass is from shaft mines. The national average for coal produced in strip mines increased from 4 percent in 1940 to 29 percent in 1972. The Donbass is a high cost producer of coal (seams are much thinner than in the fields of Asia) but this is partly compensated for by a location closer to the market. The leading coal markets in the Soviet Union are eastern Ukraine, the Urals and the Central Industrial Region. Only the Ukraine produces all its own coal needs.

Kuzbass – The Kuznetsk Basin, the Kuzbass, developed as a major project of the 5-year plans to supply the steel industry of the Urals with coking coal, is the second most important coal field and the leading field in terms of proved coking coal reserves. Originally iron ore for the new Kuzbass steel industry was to be return cargo, making the 2 000 km long rail haul feasible. However, coking coal was discovered closer to the Urals and iron ore closer to the Kuzbass, turning the simple Ural-Kuzbass-combine into a much more complex pattern. Investment per ton increase of coal output in the Kuzbass is $^2/_3$ of that in the Donbass. The Kizel field and other mines in the Urals supply more than half the coal needs of the Urals, but coking coal must be hauled from the Kuzbass and Karaganda, the third largest coal field. The high ash content of coals from this field makes it necessary to mix them with higher grade coals for metallurgical use.

Other fields – The Pechora Basin in the northeastern corner of European Russia produce good coking coals but production costs are almost as high as in the Donbass, which restricts their use to the European North and Northwest. The Central Industrial Region gets half its coal from the local Moscow Basin but higher grade coals must be brought in from other fields, primarily from the Donbass.

The cheapest coal is produced in the East where investment cost per ton increase of coal output is only some 40 percent of that in the Donbass. The Irkutsk Basin and the Bureya Basin are the major producing fields in the East. In addition there are many mines of local importance. However, much remains to be known about the composition of the vast coal reserves of northcentral Siberia, especially the tremendous Tunguska and Lena Basins and the Kansk-Achinsk Basin (brown coal).

In 1974, Japan and the USSR agreed on Japanese credits for the development of a deposit of coking coal in the Chulman area of southern Yakutia. A north-southerly branch of the Trans-Siberian Railroad will be built for the shipment of coal to Japan by way of the terminal Vostochny near Nachodka east of Vladivostok.

Asia

China – China has long been known to have large coal reserves but quantities listed in international publications are still based on estimates made before the First World War. Coal production has increased rapidly after the Revolution and China now ranks among the three leading producers. Most of the coal is located in the interior provinces of Shansi, Shensi, and Kansu of northern China, remote from centers of population and industry. Both coal and anthracite are mined in the Hwang Ho Valley. The output of the Kailan field in Hopei and other mines on the North China Plain is of greater importance. These fields are closer to the metropolitan cities, the main railroad, the Grand Canal, and the ports. The major field in Liaoning (Manchuria) is at Fushun, one of the world's largest strip mines. One seam in this field is 125 m, the thickest seam recorded anywhere. For coking, this coal is blended with coal from Penchihu east of Mukden. It supplies the iron and steel center at Anshan.

After the Revolution mines in southern China have been of greater relative importance. The mines in Kiangsi supply the steel industry of Wuhan. Other mines are worked in Hunan and Szechwan.

Japan – Coal accounted for 52 percent of Japan's energy balance in 1952 and oil for 12 percent. In the beginning of the 1960s oil surpassed coal, which in 1965 accounted for 27 percent as against 58 percent for oil. Between 1956 and 1965 Japan closed ¾ of her coal mines.

The first coal field of any importance was in northern Kyushu, where the heavy industry based on the Chikuho coal created the fourth largest of the Japanese million conurbations, Kitakyushu (= northern Kyushu). The first modern steelworks of Japan was built here at Yawata in 1901. The thin seams of the strongly faulted Chikuho-field can be worked only at high cost. In addition, the coal of this field has a high ash content. It has been used primarily in thermo-electric power plants along the coast. The city of Tagawa lost one-fourth of its population during the coal crisis.

Thanks to large fluctuations in the runoff and relatively small dam capacity, which is a result of its Alpine topography, the electric utilities of Japan, long dominated by water power, traditionally had a proportionately very large stand-by capacity in coal-fired power plants. During her remarkable manufacturing expansion in the 1960s and 70s Japan increasingly became dependent on oil-fired power plants.

From the beginning of the 1960s Hokkaido has reported larger coal production than Kyushu. The coal fields are at Kushiro on the south-east coast, at

Tab. 7 – Japan's Coal Supply (million tons)

	1970	1973
Imports:	50.2	56.9
from: US	25.2	16.5
Australia	16.5	24.9
USSR	2.8	2.8
Canada	3.4	10.4
Other	2.2	2.2
Production	39.7	23.1

Comments: The modernized Japanese coal mining industry can be expected to increase its production considerably from 1974. Imports from the USSR and South Africa will increase when large Japanese investments in mines and infrastructure have resulted in deliveries.

Tempoku-Rumoi on the west coast but primarily at Ishikari northeast of Sapporo. Mines are usually controlled by large groups like Mitsui and Sumitomo.

The third Japanese coal field, Joban, on the east coast of Honshu, has the advantage of being near Tokyo. During the coal crisis many coal miners were retrained for the tourist center of Joban.

Japan has long imported most of its coking coal, in the interwar period from southern Manchuria and northern Korea, and after the Second World War from the United States and later also from Australia and Canada. The dramatic expansion of Japan's steel industry has turned the country into a major market for coking coal.

Other Asia – North Korea, South Korea, Taiwan and North Vietnam have coal mines of importance for the national industry. The main coal fields of India are in Bihar and West Bengal along the Damodar River. Only small parts of the large reserves are suited to coking but the coal is readily accessible by rail to Calcutta and the heavy industry of Jamshedpur. It is mined in open pits and shaft mines. The large lignite deposit near Madras is being exploited on a limited scale. Coal occurs in a number of places in the eastern part of the country. In western Asia, Turkey has a coal field with coking coal from Zonguldak to Amasra on the Black Sea coast.

Southern Hemisphere

The Southern Hemisphere was long considered to have relatively small coal reserves. Prospecting in recent decades has changed the situation. Australia and South Africa dominate in production. Reserves are large also in Rhodesia and Zambia.

Australia – The Sydney Basin in Australia, with coal formations reaching the coast north and south of the city, has long dominated the coal output. Newcastle at the mouth of the Hunter Valley and Port Kembla in the southern part of Wollongong, both oriented to the coal mines, are the two dominating centers of heavy industry in Australia. In the early part of this century Australia had exports of bunker coal from Newcastle, but the modern expansion of Australian coal exports from the late 1950s almost exclusively concerns coking coal, primarily to Japan but from about 1970 also to Europe. Domestic demand for coal to feed huge thermo-electric power plants is another expanding market.

Since the end of the 1960s the vast deposits in the Owen Basin in Queensland are being exploited on a large scale. Population density is very low in this semidesert, which earlier was devoted to cattle raising, and therefore demands for relatively expensive conservation measures can hardly carry the same weight as in more densely populated strip mining districts (Appalachia, Illinois-Indiana or Montana-Wyoming). The small harbor of Gladstone in a short time was turned into one of the largest bulk ports of Australia. Cheap coal attracted the world's largest alumina plant and a large thermo-electric power plant to the port area which further increased the flows of bulk cargo (bauxite in and alumina out). In 1972 another coal harbor was opened at Hay Point north of Gladstone.

Coal was mined early at Ipswich upriver from Brisbane but these shaft mines are of only local importance. Small coal mines are also found at Leigh Creek, South Australia, and at Collie, Western Australia. The large lignite field in the Latrobe Valley of Gippsland, southwest of Melbourne, was developed in the 1920s to supply the state of Victoria with electricity.

South Africa — Coal plays a more prominent role in the energy balance of South Africa than in that of any other industrial nation, primarily because of exceptionally low production costs. The pit price for coal in South Africa is usually one-third the low American price. Seams are horizontal and close to the surface; the deepest mine is less than 250m. As a rule the mining walls are accessible by tunnel and coal can be transported to the surface by conveyor belt. The room-and-pillar system is used in the mines. Low wages paid to black workers have slowed mechanization and provided more jobs than would be normal in a European country. In spite of this, productivity per man and shift is twice the European average. Almost five-sixth of the employees are blacks of whom 40 percent are from Mozambique.

The largest coal reserves are found in eastern Transvaal with the most important mines around Witbank. This coal formation stretches south into Natal where coal is mined south and east of Newcastle. The Witbank-district and the mines along the Vaal River, straddling the Transvaal-Oranje Free State border, with mines north and south of the steel city of Vereeniging, have a favorable location in relation to the gold mines which consume much of the generated electricity. The state-owned Electricity Supply Commission (ESCOM) accounts for some two-thirds of the produced electricity. The large power plants of ESCOM have lower production costs than thermo-electric power stations in any other country, as low as those of hydro-electric power stations in other parts of the world.

During the post-World War II-period South Africa has been the only country with a facility for making synthetic gasoline from coal. South African Coal, Oil and Gas Corporation (SASOL) in 1955 built a plant at Sasolburg 22 km south of Vereeniging. It is based on the German Fischer-Tropsch-principle. During the Second World War synthetic gasoline accounted for much of Germany's needs but production costs were high. The coal mine at Sasolburg is completely mechanized and produces four times as much coal per man and shift as the average. But in spite of record low coal prices the plant is not profitable. It has accounted for some 4 percent of South Africa's needs. However, the drastically increased oil prices in 1973/74 and the embargo on the export of oil to South Africa and other countries have drawn the attention of the world to Sasolburg.

Rhodesia — South Africa's exports of coal will increase with the opening of the coal terminal at Richard's Bay on the Indian Ocean. It is connected by pipeline with a new coal field. It is the largest infrastructural investment in South Africa, almost 2 billion dollars. The technique with coal transport in a slurry through pipeline was used in Ohio in the 1950s, but when unit trains were organized on existing railroads they turned out to be competitive. During World War II, coal slurry was piped in Germany, eg from the Ruhr to the Salzgitter steelworks.

Good coking coals have been mined since 1903 in Rhodesia at Wankie 110 km southwest of the Victoria Falls. Mining takes place at depths of 25–60 m in seams of 3–10 m and the coal is hauled through tunnels. Costs are low but said to be somewhat higher than in South Africa. Rhodesia as well as Zambia and Malawi, has tremendous reserves of low-grade coal with an ash content of 20–35 percent.

Latin America — Brazil's coal fields are in the southern states of Santa Catarina and Rio Grande do Sul. With high ash and sulfur contents Brazilian coals must be processed and mixed with imported coals to produce coke at the ovens of the steel mills. Coal mining south of Concepcion in Chile faces similar problems. The mines at Coronel and Lota are partly worked under the sea. They supply the steelworks at Concepcion.

North of the Equator Latin America has some coal production in Colombia and Mexico but the whole area in 1970 produced only 9 million tons. Latin America is poor in coal resources. Colombia produces coal at Cali in the Cauca Valley and at Pas del Rio northeast of Bogotá. The latter mine supplies a new steelworks. Among several known coal deposits in Mexico the most important is near the US border between Piedras Negras, Sabinas, and Lampazos. The mines at Sabinas are on a major railroad and supply the steel industry of Monterey.

Outlook for the coal industry

Estimates of coal reserves may be more accurate than those for petroleum and natural gas but at best they are indications of the relative order of magnitude. Any estimate shows coal reserves to be enormous in relation to current production. After recent revisions, Asia, especially Siberia, is shown far ahead of North America, primarily the United States, in total reserves. Europe ranks third.

Recent trends in coal mining include 1) the sharp upsurge in Soviet outputs until the early 1960s, followed by a slowing down when the country decided to emphasize cheaper forms of energy (oil, natural gas), 2) the decline in American production until 1961 while most coal users switched to petroleum or natural gas, followed by a conspicuous expansion caused by rapid growth in the two remaining markets (thermoelectric power stations and coke ovens), 3) a rapid decline of coal mining in Western Europe including Britain in spite of Government subsidies motivated on social grounds because of unfavorable geology and consequent low outputs per man-day (US 18 tons, Australia 10, Germany 3.7, Belgium 2.6 and United Kingdom 2.2 in 1970). International trade in coal is mostly for coking coal. As a fuel in power plants, coal is competitive only near the mine or where cheap transport can be provided (barges, unit trains). World production of coal continues to grow, now mostly based on its remaining two markets, both of which are rapid-growth markets. Coking coal is expected to become a commodity in short supply within the next few decades.

In the longer perspective, coal is seen as a raw material for synthetic gas and gasoline when the cheaper oil and natural gas deposits have been exhausted. The events of 1973/74 may have shortened that perspective considerably which may lead to a more rapid growth rate for the coal curve in the decades to come.

Petroleum – Oil and Natural Gas

Oil and natural gas often occur together and have much in common in nature, exploration and use. Traditionally petroleum includes solid (asphalt), liquid and gaseous hydrocarbons. A striking feature of the world petroleum supply is its concentration in areas remote from the manufacturing belts and densely populated regions. By the time petroleum, literally rock oil, came into major use, the leading manufacturing belts had long been established near the coal fields. A major share of the world's petroleum production moves to these older industrial regions. Only relatively late, primarily after the second World War, did petroleum become the basis for new industrial complexes with oil refineries, petrochemical plants, thermo-electric power stations, steelworks, alumina plants, fertilizer factories and so on. The petroleum complexes are primarily located in major ports of the old industrial regions with some also in the shipping ports.

Japan has built completely new ports and manufacturing districts to benefit from the advantages of the new industrial complexes: energy economy, industrial linkages (the output of one factory is the input of another), deep quay location, common entrance channels and suitable industrial land (dredged masses are used as land fill).

The tremendous commodity flows resulting from the disparate patterns of production, refining and consumption of petroleum have long dominated the

Tab. 8 – International Trade in Crude Oil 1971 (SITC 331) (billion US dollars)

Imports		Exports	
Japan	3.04	Saudi Arabia	3.35
United Kingdom	2.33	Kuwait	2.95
Italy	2.29	Libya	2.68
Germany (W)	2.21	Iran	2.00
France	2.17	Venezuela	1.93
United States	1.88	Iraq	1.46
Netherlands	1.55	Nigeria	1.33
Spain	0.66	Canada	0.78
Belgium	0.65	Indonesia	0.44
World Total*	22.81	World Total*	16.70 (estimated)

* Socialist Countries excluded.
Source: YITS 1972–73.

Tab. 9 – International Trade in Oil Products 1971 (SITC 332) (billion US dollars)

Imports		Exports	
United States	1.44	Netherlands	1.11
Germany (W)	1.00	Venezuela	0.89
Japan	0.58	Italy	0.79
United Kingdom	0.56	Neth. Antilles	0.62
Sweden	0.51	United Kingdom	0.49
Switzerland	0.31	United States	0.47
Denmark	0.27	Singapore	0.45
France	0.27	Germany (W)	0.39
Netherlands	0.26	Trinidad	0.36
World Total*	8.17	World Total*	7.70 (estimated)

* Socialist Countries excluded.
Source: YITS 1972–73.

world transport map measured by tonnage, but are also leading by value, tables 8–9.

Petroleum fields are associated with sedimentary basins. Both crude oil and natural gas occur in porous rocks, primarily sandstone and limestone, covered by an impervious layer, a cap rock, frequently shale. To pool and hold the migrating oil within a reservoir the sedimentary layers must contain a trap-structure: an anticline, a salt dome, or a fault. The trap blocks the movement of the oil which accumulates in the upper part of the formation.

It is now generally held that petroleum of varying densities and of varying mixtures originated in enormous quantities of marine life, especially plankton,

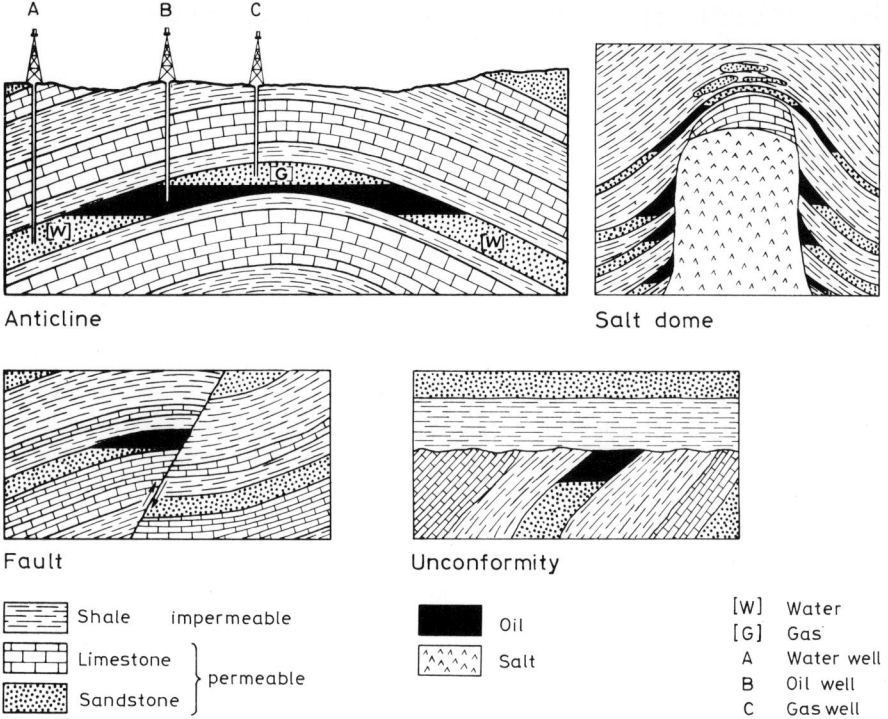

Fig. 13 — A selection of common petroleum traps.

that were buried in deep silts on shallow ocean bottoms. This process still takes place and oil has been found in formations of all geological periods from Cambrian onwards but primarily in the more recent sediments. For the formation of coal, Carboniferous was an advantageous period, but no such peak period has been indicated for petroleum. However, more than half the oil is obtained from Tertiary deposits and very little from Cambrian and Ordovician sediments.

Oil

Oil has been found in all continents, including continental shelves. Large areas of the world have sedimentary structures capable of holding petroleum. Several early oil fields were easy to find; seepages had been known for hundreds or thousands of years. The Zoroastrians (Parsees) early in the Christian era reverenced eternal fires in southwest Asia, eg in Persia and at Baku. Other oil carrying structures can be identified in the pattern of surface outcrops of rocks. But in many areas, structures are buried deep beneath the surface. In such cases, geophysical exploration is necessary to identify potential oil-bearing structures.

Gravimeters and magnetometers are flown over the area to collect data for detailed gravity and magnetic maps.

Teams are sent to interesting areas to make seismic surveys, which allow a detailed picture of the underlying structures. However, the only way of finding out whether petroleum is actually present in promising structures is by drilling a test well, a wildcat. Additional wells have to be drilled to outline and develop the field. Exploring for petroleum must often be carried out in the face of great physical difficulties: in the swamp of a tropical rain forest, in a desert, on the tundra or from the bottom of shallow seas. Several million dollars are often spent before the first barrel of oil is produced in a new field. Risks can best be carried by large organizations; oil groups are among the largest industrial corporations of the world. But the time is long past when the 'Seven Sisters' formed an oligopoly in the international oil market. In the North Sea and other new oil producing districts hundreds of companies are trying their luck.

In densely populated areas, the land of an oil field may have many owners while in deserts and other sparsely populated regions the land often has only one owner, as a rule the state. The optimum spacing of wells can then easily be arranged. Wherever a border line runs through an oil field, operators will flock

Tab. 10 – The Eleven Largest Oil Groups in the World in 1972

Company	Sales (million dollars)	Employment	Rank**
1 Exxon, New York	20310	141000	2
2 Royal Dutch/Shell, Ha/Lo*	14060	174000	(1)
3 Mobil Oil, New York	9166	75400	7
4 Texaco, New York	8693	76500	8
5 Gulf Oil, Pittsburgh	6243	57500	11
6 Stand Oil of Calif, S F	5829	41500	12
7 British Petroleum, London	5712	70000	(4)
8 Stand Oil (Ind.), Chicago	4503	46600	15
9 Continental Oil, Stamford	3415	38100	24
10 Atlantic Richfield, L A	3321	27800	25
11 Cie Francaise des Pétroles, Paris	2806	24000	(27)

* Shell has headquarters in the Hague and London.
** Rank refers to two lists, one for manufacturing and mining corporations in the United States and the other ( ) for corporations outside the United States.

In addition to the eight listed American oil companies, another five had sales in excess of one billion dollars in 1972: Phillips, Occidental (Oxy), Ashland, Getty and Marathon. Outside the United States were eight such companies in addition to the three on the list: ENI (Agip) Italy, Elf-Erap France, Petrofina Belgium, Petrobrás Brasil, Idemitsu Kosan Japan, Pemex Mexico, Gelsenberg Germany, and Maruzen Japan.

After the oil crisis 1973 the German VEBA and Gelsenberg groups were merged. Government ownership in the VEBA-group exceeds 40 percent. By sales (25 billion DM in 1974) or employment the new group would rank high in the list.

to that line to tap the common oil pool. An extreme case is Signal Hill, Los Angeles, developed in the 1920s. Here operators controlled only one or a few city lots and drilling rigs were closely spaced like trees in a wood. Yields per well will be low in such fields. But the structure per se is also important for the yields per well. The large and regular structures of the Persian Gulf area are tapped by relatively few wells. A well in this region on the average produces several hundred times as much oil as the average American well. Saudi Arabia is known for her extremely low production costs, less than 5 cents a barrel in the early 1970s, and the United States for her high costs, almost 3.00 dollars a barrel at the same time in marginal wells. The American marginal costs were also the global marginal costs.

Three major types of crude oils are distinguished, varying widely in their gasoline yields under straight-run distillation. The paraffin-base crudes have the highest yields and the asphalt-base crudes the lowest, some yielding no gasoline at all, with the naphthenic-base crudes in an intermediary position. Generally crudes are a combination of the three types. As the art of manipulation and chemical transformation has made great strides, the original character of the crude has lost much of its importance as a determinant of yields of specific products. However, some trade flows, seemingly irrational, may still primarily be explained by differing characteristics of crude from field to field.

Potential oil supplies of the world are a highly controversial topic on which experts disagree widely. Newspapers have contained alarming reports about the short remaining life-time of the petroleum industry almost from the time the first well was drilled. The high cost of exploration prevents the industry from looking more than a few decades ahead, but proved reserves have almost always been bigger at the end of a year than at the beginning. However, growth of consumption at the rates of the oil industry's first century would undoubtedly within a relatively short time exhaust reserves however large they may be, even if the hydrocarbons of oil shale and coal were included. New sources of energy must be found to allow remaining reserves of petroleum to be used as a raw material for the petro-chemical industry.

Oil refining – Refining means separation of the hydrocarbons mixed in the crude oil. The simplest process – the only one used before 1913 – is fractional distillation. The various hydrocarbons have different boiling points. When crude, heated to $315°-370°C$, is fed into a distillation tower, its components rise as vapor to different levels of the tower, condense, and are drawn off.

The straight-run products may be treated further in other processes such as cracking, catalytic cracking, polymerization, hydrogenation, and alkylation. Through the cracking processes more of the lighter fractions, gasoline, kerosene and distillate fuel oil, are obtained and less residual fuel oil. In addition, the cracked gasoline has a higher octane rating than straight-run gasoline. However, some light hydrocarbons may be too volatile for use as gasoline. They are combined into larger and more stable molecules by polymerization, hydrogenation or alkylation. Lubricants, made from medium and heavy distillates, are also processed chemically before being blended into numerous lubricating oils and greases. A large refinery may be a complicated structure with a certain economy of scale but it may also be a simple distillation plant. Two refineries with equal throughput may vary widely in size measured by employment or invested capital. It has been possible to design small and simplified refineries suited for small markets, e. g. tropical countries.

Transportation — By tonnage, petroleum, including natural gas, dominates all trade flows in the world. The means of transport are pipelines, tankers and barges, tank trucks and railway tank cars. From simple beginnings at the end of the 19th century the world fleet of tankers and barges and the net of pipelines have been expanded to tremendous dimensions. The economies of scale are very large in the transport of petroleum. The first oil tanker, the steam-powered Zoroaster, built for the Nobel Brothers of Baku at the Norrköping shipyard in 1878, was small, 249 dwt, compared with the largest ship afloat in early 1973, the 483 650-ton-tanker Globtik Tokyo built at the Kure shipyard in Japan. The largest tankers of the 1970s cannot enter the North Sea or pass through the Strait of Malacca. A special reloading port has been constructed at Bantry Bay, Ireland. Even with the reloading costs included, giant tankers are economic in operation on long runs, e. g. the Persian Gulf-Western Europe. Where the two are alternatives, pipelines cannot compete with tankers. Pipelines serve as feeders to shipping routes, from inland oil-fields to ports and from ports to inland refineries.

Location of oil refineries — Basically, oil refineries are located near the market or in the oil shipping ports. Since freight rates for dirty oil (crude and heavy fuel oil) is some 10% lower than for clean oil (gasoline and other light fractions) a location near the market is advantageous as long as the output of the refinery matches the local demand; no back-hauling is then necessary. As mentioned, the yield of the straight distillation can be changed, but only at a cost. Installations for cracking and so on are heavy investments.

A location in the oil shipping port allows products to be sent in various directions and usually also meets a large demand for bunker oil from tankers that fill up on bunkers as close as possible to the oil field. Strategic considerations also influence the location of oil refineries. In an emergency it is relatively easy to find new sources of crude oil for refineries located in industrial countries, but difficult to supply the markets with products from alternative refineries if many of them are located in shut-off shipping ports. Political stability of the market regions and instability of many source areas weigh heavily in favor of the market area as location for oil refineries.

North America

The United States – America was long the world's largest producer of oil, a position held almost all years after the first well was drilled in 1859. The American share has been declining since 1948 when it was over 60%, a ratio maintained since 1906, and was only 21% in 1970, in spite of record production. The American output declined from 475 million tons in 1970 to 432 million in 1974 and even less in 1975.

The Appalachian fields were the first to be developed. The historic well at Titusville, northwestern Pennsylvania, was drilled in 1859 and subsequently other fields were developed in western Pennsylvania, southwestern New York, southeastern Ohio, West Virginia and eastern Kentucky. Cleveland, Ohio, became the first managerial center of the rapidly expanding petroleum industry.[10] Appalachian oil is high-grade and light with a paraffin base, no sulfur, and well-known for its high-grade lubricants. But production by modern standards has remained small and slowly declining.

The Middle West has two more areas of oil production, one stretching from northwestern Ohio and northeastern Indiana into southern and central Michigan. In the southern part of this district shallow wells were drilled in the 1880s and early in this century but output declined sharply after a few years. Deeper wells were drilled in Michigan in the 1920s. But few wells are now pumping in these fields. The other area is in southern Illinois, southwestern Indiana and western Kentucky. Fields were first developed here in the early 1900s.

The original oil region in the United States, partly coinciding with the dominant bituminous coal area, was eclipsed at the turn of the century by enormous finds of oil made in areas peripheral to the northeastern Manufacturing Belt, in Texas and adjacent states and in California. The large mid-continent region, stretching from Kansas through Oklahoma and northern and western Texas into southeastern New Mexico, has hundreds of oil fields, the first of which went into operation in the 1890s. It has long ranked as the leading American oil region, now accounting for over a third of the national output. The Gulf coastal region with many fields in the coastal plains of Texas and Louisiana and reaching into Arkansas, Mississippi and Alabama is only slightly smaller in output. Since the 1950s, new offshore districts have been added to this region. A typical feature of the Gulf coastal region are the many salt domes which act as oil traps.

The third largest oil region is southern California with three major districts: the Los Angeles basin, the San Joaquin Valley and the coastal district. The first well was drilled in 1887, but large-scale production had to await the opening of the Panama Canal in 1914. At the turn of the century the American west coast was still a sparsely populated part of the country. The subsequent rapid population increase has turned California into the most populous state of the federation in which, for lack of competing coal fields, petroleum completely dominates the energy balance.

The large hydroelectric power stations of the region only contribute small shares. The continuation of the Gulf of California, some 240 km, on the American side of the border, the Imperial Valley, is underlain by a giant salt sea under remarkably high temperatures and pressures. Difficulties in finding materials that function under such temperatures and pressures have caused this potential energy source to remain untapped. It could supply the whole of southwestern United States with energy for a couple of hundred years. The geothermic sources at Sonoma north of San Francisco are expected to have a capacity of 900 MW in 1977.

In spite of large local production, California is now a deficit area for petroleum. The balance of oil is received from Alaska (1977) and the balance of natural gas is piped in from primarily the Midcontinent region and the Rocky Mountains, but also from Canada.

The many small fields of the Rocky Mountains and the northern Great Plains together produce more oil than California. Wells are deep and, as a rule, of recent origin. They yield high quality light oils. These fields continue on the Canadian side of the border.

In the extreme north of the plains area, the Arctic Slope of Alaska and Canada, vast deposits of oil were discovered in 1968. Inaccessibility has delayed the exploitation of these fields. The icebreaking 115 000-tons-tanker Manhattan in 1969 made a test run through the ice of the Northwest Passage from the American east coast to the Prudhoe Bay oil field, the largest ever found in the United States. Tanker shipping in Arctic waters turned out to be technically possible. If it can be made economically feasible and ecologically defensible the strategic location of the Arctic Ocean will come to its right. Prudhoe Bay, the delta of the Mackenzie River and the Beaufort Sea are almost equidistant from Philadelphia, Rotterdam and Tokyo, the three large oil markets of the world. The environmental threat from tanker shipping is very great in the Arctic surroundings; many problems have to be solved before regular shipping can start with icebreaking or submarine tankers.

The first outlet for the Alaskan oil will be the 1 270 km pipe line (125 cm, 48 in.) between Prudhoe Bay and the little town of Valdez on the south coast. The first oil will flow in 1977 and the final capacity will be 60 million tons a year. The 1 000 inhabitants of the town heard of the pipe line in 1969, but not until April 1974 had the necessary laws passed congress after comprehensive investigations and one of the most intense environmental debates in American

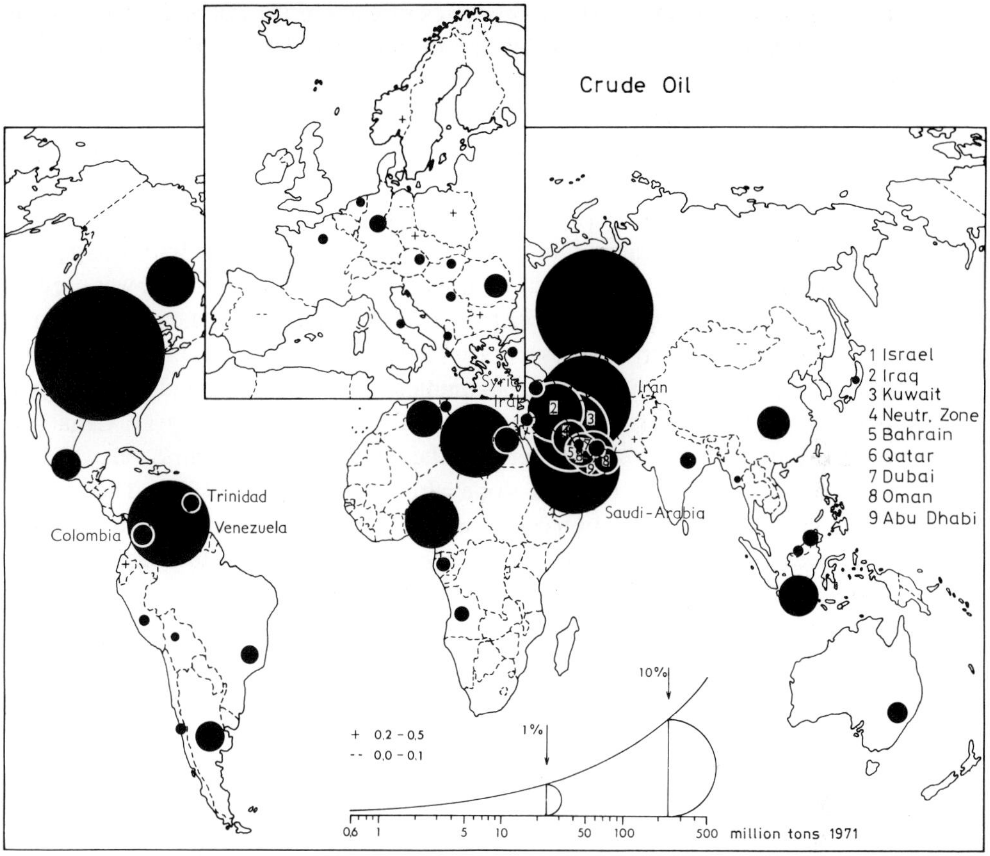

Fossil Fuels 49

Fig. 14 – Map of world oil production before the Deluge. It will take many years before the map again has settled down to, from year to year, reasonably stable relations. In 1975, some countries had a rapidly expanding production, others a declining output.

history. A pipe line across Canadian territory down to the system at Edmonton has also been considered.

Canada – The Canadian oil production is primarily located in the prairie provinces, Alberta and Saskatchewan. The first field of importance was Turner Valley (1936). The large expansion started in 1947 with the discovery of Leduc near Edmonton. Since the Canadian oil consumption is strongly concentrated to the southeast and large American markets are closer than the domestic ones to the producing fields, Canada in the early 1970s exported as much oil as it imported, some 25 million tons each. The long pipe line between Edmonton and Sarnia, Ontario, partly runs over American territory. The pipe line supplying the refineries of Montreal with oil from Venezuela has its tidewater terminal at Portland, Maine. After the oil crisis of 1973/74 a continuation of the Sarnia line to Montreal was started. If the refineries of eastern Canada should be cut off from foreign supplies the capacity of the line could be increased from 12.5 million tons a year to 25 million tons. Plans exist for constructing a pipeline north of the Great Lakes entirely on Canadian territory.

Alberta in 1974 produced 83% of Canada's oil or 72 million tons a year. The price at the well in one year until October 1974 increased from $ 4 a barrel to $ 6.50. Of the price increase, two-thirds went to the Provincial government. In addition the exports carried an export tax of several dollars a barrel to get the export price in line with the fluctuating world market price. This tax was used by the federal government in Ottawa to subsidize oil imports into eastern Canada.

Canada has very large oil reserves in the tar sands of Lake Athabasca, known for a long time. Exploitation on a commercial scale was started in 1968 by a plant at Fort McMurray with a capacity of 2.5 million tons a year. Canada is expected to get a substitute here in the 1980s for the prairie oil which is expected to have a downward slope after a peak about 1976. The remote location and high exploitation cost of the tar sand oil is a handicap that only high oil prices can compensate for. The pipe lines from coast to coast may also distribute oil from off-shore fields that the Canadians hope to find on their shelves in the Atlantic Ocean.

Mexico played a major part in the world production of oil for some years in the early part of this century. The coastal plain in the hinterland of Tampico experienced an oil boom after 1901 that brought Mexico to the second place among the oil producers of the world with a maximum of 28 million tons in 1921. Large deposits near the surface and near the coast attracted competing foreign companies. But the rapid rise of Mexico's oil production was followed by a drastic decline due to rapid exhaustion of oil wells and lack of development of new fields following increased taxes and government restrictions. The friction between government and oil companies in 1938 led to expropriation of foreign-

→

Fig. 15 – The production of oil comes close to having had a consistent exponential growth during its more than 100 years long history. In the semilogarithmic diagram it is represented by an almost straight line. The decline in growth rate before 1973 was small, indeed, but after 1973 not only the growth rate but total production declined. The first decades of oil history showed a very rapid increase. The oil curve is far from the typical logistic curve with its slow and hesitant initial period.

On all diagrams in the series, annual growth rates ('compound interest' rates, geometric progression, exponential growth) is indicated between two years marked by arrows. The doubling time for oil production has almost consistently been somewhat less than 10 years (7.2% annual increase).

Even without the oil crisis, the growth rate of oil production would have declined in the 1970s. By then, oil had conquered all traditional coal markets except two, in which coal (and brown coal) were still competitive near the coal fields. With a much higher percentage of the total energy market, oil would have adapted itself to the lower growth rate of total energy, fig. 5.

Fossil Fuels 51

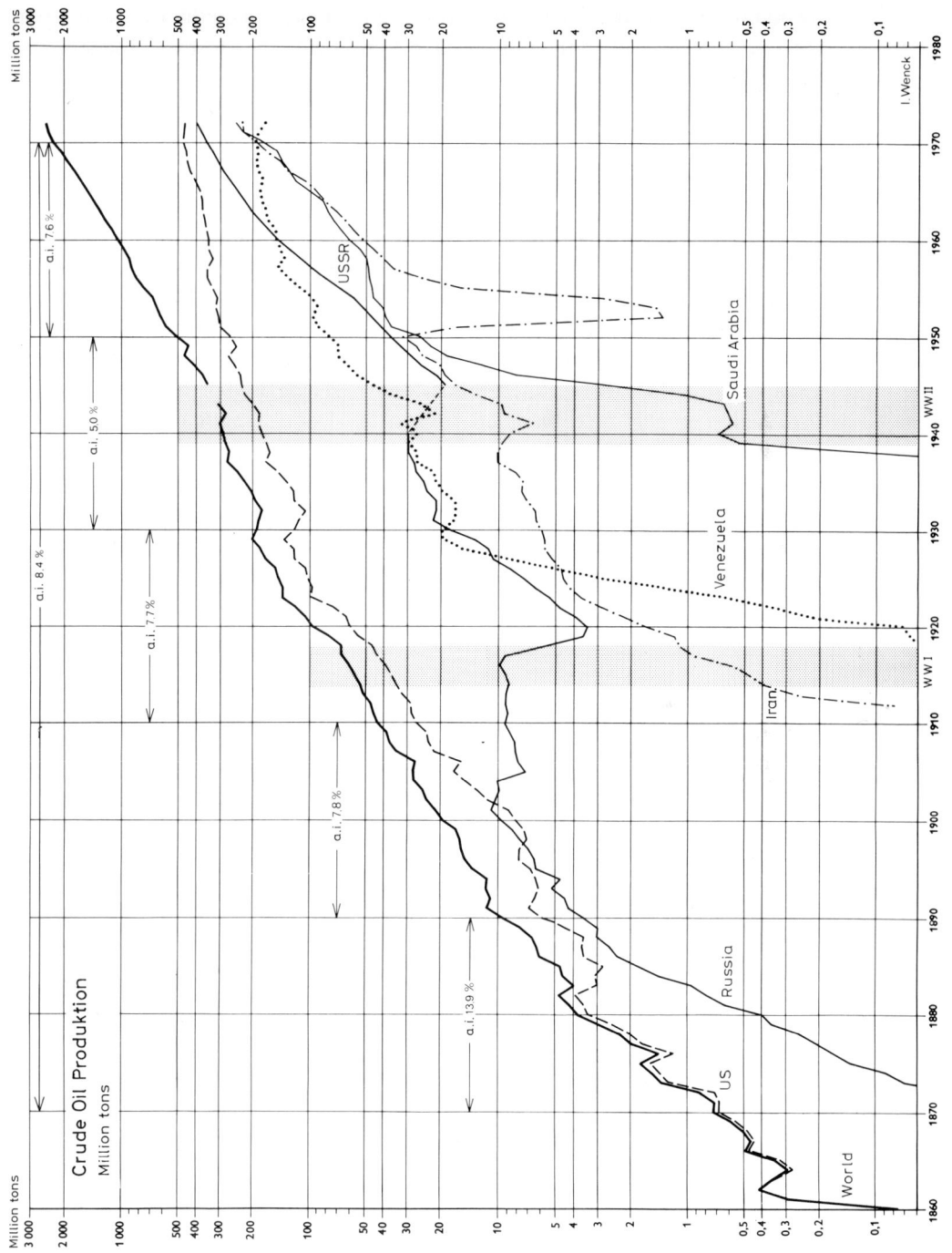

owned oil properties. The oil companies had then long seen the handwriting on the wall and had moved their interests to Venezuela. The Mexican production at the end of the 1930s was 5–6 million tons.

Under the state-owned Petróleos Mexicanos (PEMEX) the many fields from north of Tampico to the Isthmus of Tehuantepec and in the adjacent offshore areas have slowly expanded production, now primarily for the domestic market. New exploratory drilling is undertaken with the help of foreign oil companies working under contract. Domestic oil consumption is encouraged by the government; the alternative fuel, wood or charcoal, incurs deforestation which creates many problems of soil erosion and water conservation in tropical environments.

New large oil finds were made in the mid-1970s in Mexico. The Reforma-field in marshy land south of Villahermosa on the border between Tabasco and Chiapas has by foreign oil experts been compared with the North Slope of Alaska. Production in this field went up from a level of 3.5 million tons a year in December 1973 to 15 million tons in March 1975. This should be compared with a total oil production level of 35 million tons a year for Mexico at that time. Other promising finds have later been made near the Contaxtla River in Veracruz and at Chac in Campeche. Mexico in 1975 was becoming a net exporter of oil. Pemex plans to produce over 100 million tons in the early 1980s of which over half will be exported.

South America

In South America all countries except Paraguay and Uruguay produce oil. Some deposits were known to the early Spanish seafarers who used petroleum for calking their ships. The vast asphalt lake in Trinidad was recorded by Sir Francis Drake.

Trinidad once was the leading oil producer in the British Empire. It is still a major producer in Latin America. In 1974 rich oil and gas fields were found off Trinidad's east coast which again have pointed the production curve upwards.

Venezuela for decades before 1960 ranked second only to the United States as an oil producer. After 1960 the Venezuelan oil production stagnated while the Soviet output expanded rapidly and in 1975 was more than four times as large, even exceeding the American in size.

The first modern oil well was drilled in 1914 at Mene Grande and three years later in the La Rosa field, both near Lake Maracaibo. Prospecting was stepped up in 1922 when one La Rosa well started blowing wild for an estimated 5 million tons a year, which called the world's attention to the tremendous production potential of the Lake Maracaibo area. The active prospecting that followed in the Lake Maracaibo region and in the llanos of the Orinoco basin of eastern Venezuela coincided with one of the recurrent American debates about the country soon running out of domestic oil reserves.

The Big Three, Standard Oil of New Jersey (Creole), Shell, and Gulf (Mene Grande), have dominated Venezuelan oil production since the late 1920s and in 1970 accounted for over 80 percent of the crude output.

When Shell in 1917 developed the first large oil field in the Lake Maracaibo region it was necessary to find a transshipment point outside the lake area. The shifting sand banks at the entrance of the lake prevented the passage of oceangoing tankers. Willemstad on Curaçao was chosen for the oil terminal and refinery. Venezuela was still a small out-of-the-way republic with a primitive agropastoral economy, known for its unstable governments. Points on the mainland must have seemed less attractive than the Dutch island for a large investment by the joint British/Dutch corporation. In 1928 Esso chose another of the Netherlands Antilles, Aruba, as the site of its refinery.

The long conflict that led to the nationalization of the oil fields in Mexico 1938 forced the oil companies to look for alternative oil suppliers. The Second World War led to a tremendous increase in oil consumption in the United States that could not be met with increased domestic production. Venezuela saw its production go up from 15 million tons in 1928 to 31 million in 1939 and 70 million in 1948.

After the Second World War, Shell und Esso built oil terminals and large refineries on the west coast of the Paraguaná Peninsula at Punta Cardón (1949) and Amuay Bay (1950), following government legislation to stimulate petroleum refining within Venezuela's borders. In 1956 Lake Maracaibo was opened to ocean-going tankers when a deepwater channel was completed at the mouth of the lake, subsequently deepened and extended through the Straits of Maracaibo. Before the channel was built, crude oil was transported by shallow-draft vessels from a number of small terminals along the lake shore to the large refineries and terminals on Aruba and Curaçao but the deep-water approach allowed loading to be concentrated to a few big lake terminals, like Puerto Miranda (Shell) and La Salina (Esso).

As there are no surface indications of oil in eastern Venezuela, oil production in this region had to await the development of the new science of geophysics. After many years of exploratory drilling oil was found in 1928 near Caripito in the eastern llanos. In 1937 Gulf found oil in the rich Oficina field in the central llanos. Oil from this field was sent to Puerto La Cruz, which became a major oil terminal. Since 1947 an oil district has been developed in the western part of the llanos in the Barinas basin. A pipeline was built to the El Palito refinery west of Puerto Cabello.

During the oil crisis of 1973/74 French interests signed a preliminary agreement to develop the large tar sand deposits along the Orinoco River in which the recovery rate is expected to be only some 10%. The reserves are estimated to be of similar size as the Canadian.

Oil has put Venezuela on the shipping map of the world as the most important country of Latin America, far ahead of the Plata region at the peak of its grain shipping heyday. Petroleum has long accounted for over 90% of Venezuela's export value, for over 60% of the government revenue, for over 20% of the GNP, but for less than 3% of the employment. Oil money has provided Venezuela with funds for modernizing the economy of the country. Cities, ports, factories, and highways have been built but the welfare of the majority of the Venezuelan population may not yet have changed profoundly.

Venezuela's production curve shows a marked break point in 1958 with an annual increase of 10% in the preceding 15 years and less than 3% in the following 15. This may be given a political explanation. The government in power before 1958 favored foreign investments. The incoming government had a socialist and nationalist program. By holding back production it wanted to support prices and save oil for the future development of the country.

Changes in the global economic geography of oil may provide a more credible explanation. Production costs in Venezuela are much lower than in the United States. American oil companies had to ask for protection in the 1930s to survive competition from cheap Venezuelan oil. After 1958 exceedingly cheap oil from the Arabian/Persian Gulf and later from North Africa and Nigeria started to make inroads on Venezuela's traditional markets. Ever larger super tankers diminished the competitive advantage of shorter distances. In production costs, decided primarily by the geology of respective areas, Venezuela turned out to be in an intermediate position between the high-cost producer, the United States, and the low-cost producer, Saudi Arabia. The investments of the oil companies naturally went primarily to the low-cost areas. The enthusiasm of Venezuela for the creation of OPEC in 1960 is thus obvious. Venezuela had everything to gain and nothing to loose from an oil cartel. With low energy prices, other oil exporters would get an ever larger piece of the pie. With high prices, a shift to other energy forms would be speeded up, but Venezuela would in the meantime be in no worse position than the extreme low-cost producers. Venezuela would get its share of the monopoly profit.

The assets of the foreign oil companies in Venezuela were taken over by the state-owned oil company Petroven from January 1976.

Colombia has four petroleum-producing areas: the middle Magdalena Valley, the extreme southwestern corner of the Maracaibo Basin, and the lower and upper Magdalena Valley. The first two produce most of the oil. Pipe lines connect them with the major cities and with export terminals south of Cartagena (Mamonal and Coveñas).

Peru — Petroleum production in the northernmost part of the coastal desert of Peru, served by the oil port of Talara, dates back to the end of last century. The field now produces 2 million tons a year, the same as a new field off the coast.

In the jungle east of the Andes, 29 foreign companies and the state-owned Petroperu were prospecting in the early 1970s. By 1975, only Occidental and Petroperu had found oil. The pipe line across the Andes to Bayovar on the coast, built by the state at an estimated cost of more than half a billion dollars, may prove to be excessively large when ready in 1976.

Ecuador in 1972 became an oil exporter from its fields in the Amazonas region. The oil is produced by a company owned jointly by Texaco and Gulf. The 510 km pipeline over the mountains from Lago Agrio to the port of Esmeraldas has a capacity of 20 million tons a year.

Argentina's first oil was discovered at Comodoro Rivadavia on the coast of Patagonia in 1907. Other fields are located inland just east of the Andes at Plaza Huincul, Mendoza and Salta. With an output exceeding 20 million tons a year, Argentina is largely self-sufficient in oil. Like some other Latin American countries, primarily Brazil, Argentina has been reluctant to let multinational corporations participate in the exploitation of its oil, which presumably has held back the output. The sediments, from Tierra del Fuego in the south to the llanos in the north, should hold large oil reserves, more than is reflected by present output from the Andean countries.

Europe

Europe, the home of the Industrial Revolution, was well provided with coal but was long considered to be poorly equipped with petroleum. Romania was the obvious exception. The Ploesti field claims to have produced petroleum from dug wells before the first well was drilled at Titusville, Pennsylvania, in 1859. Petroleum was used as a patent medicine for men and horses of the stagecoaches running through the area.

The technique of underwater drilling, first developed in Lake Maracaibo and off the coasts of Texas-Louisiana and California, have made Europe's prospects as an oil producing region look much brighter in the mid-1970s than a couple of decades earlier. The vast shelf areas off Europe's long coast lines are much more interesting to oil prospectors than the on-shore areas. Professor Peter Odell of Rotterdam, an expert on the economic geography of oil, gives Western Europe a better chance than the United States of becoming almost self-sufficient in oil and natural gas.

Romania — In the interwar years Ploesti completely dominated the Romanian output, which during World War II became an important strategic objective. After the war, new fields have been developed at Arges, Ticleni, and Bacau. Annual production for a long time has remained stable at 13 million tons or roughly twice the prewar output.

Germany has long produced some oil from small fields; in the late 1960s the output approached 8 million tons. Austria, France, and Italy have much smaller production.

Norway — The Norwegian Ekofisk field in the North Sea is expected to reach the production level of 25 million tons a year in 1976. Since domestic consumption is less than 10 million tons, Norway will be an important oil exporter. The oil from Ekofisk is piped to Tees-side in England and the gas to Emden in Germany. The field is west of the submarine trough that extends from the Norwegian Sea around southern Norway into the Skagerak. At its deepest part the Norway Deep exceeds 700 m, a formidable obstacle to submarine pipelines. Norway has other proven oil fields in the North Sea, most of which have a common oil pool with the British fields on the other side of the median line between the coasts of the two countries, which has been agreed upon as the border line for mineral exploration. Facing the prospect that Norway may become a very large petroleum producer, the Norwegian government has stated that Norway, in order not to disrupt the Norwegian way of life, will not produce more than 90–100 million tons of petroleum a year. This may obviously lead to a conflict of interest between Norway and the rest of Western Europe. It may also mean, that the Norwegians will have to allow the British interests on the other side of the line to pump all oil and gas from the common pool in order not to exceed their target production. The answer to the interesting geopolitical question — will it be possible to hold back production in a small country under pressure from external and internal groups — will be given in the next two decades.

Shelf areas along the Norwegian west coast, north of latitude 62° N and in the Barents Sea, including the Spitzbergen Islands, have very promising structures. In the Barents Sea, Norway and the Soviet Union have conflicting interests. Norway claims that the median line principle should be applied between the Norwegian Spitzbergen and the Soviet Novaja Semlja, while the Russians claim the sector principle (a straight line from the North Pole to the Norwegian-Soviet border).[11]

United Kingdom — The British have made even larger petroleum finds than the Norwegians in the North Sea, gas in the south and primarily oil off the coast of Scotland, where Aberdeen has had an oil boom exceeding that of Stavanger in

Norway. The British expect to be self-sufficient with oil by 1980. The first oil from British fields was landed in 1975. The Irish Sea and the shelfs west of Ireland are also of great interest to petroleum prospectors.

The Soviet Union

The Soviet Union since 1953 has had a rapid expansion of its oil production with an annual increase exceeding 12%. In 1974, its output exceeded the American for the first time since the turn of the century. The Russians in 1974 produced 459 million tons, following a rising curve, and the Americans 432 million tons, along a declining curve. According to projections, Saudi Arabia should have been in a clear first place, but, for well-known reasons, it was a poor third with 408 million tons. Oil consumption in the two super countries were in quite other relations: the Soviet Union exported over 100 million tons and the United States imported more than 300 million tons.

Baku — Russia was one of the pioneers in the modern development of oil fields, utilization of tankers and the construction of pipelines. From the 1870s the fields on and off the Apsheron Peninsula in Azerbaijan were developed into one of the major oil districts in the world. Baku became a major Russian industrial center. Oil was sent by way of the Caspian Sea and the Volga river into the heartland of the country, and, later, also by way of the Baku-Batumi pipeline (built 1896–1906) to the Black Sea and the world market. At the turn of the century, Russia for some time was the world's leading oil producer. Until the late 1930s, Baku accounted for over 70% of the Russian output, but in 1972 this share had declined to less than 3%. The production of the latter year was about half the record production during the Second World War. The output of gas continued to climb.

The Soviet geologists have had to go deep for oil in Baku, 6400 m on-shore and 5485 m off-shore, and they expect to go even deeper. They pump gas or steam into the wells to keep the pressure, which means high costs. The first well in the Caspian Sea was drilled in 1949 and now $^2/_3$ is produced off-shore. Some 120 km off-shore is a 13 km oil-rig-complex, where 5000 people work in shift with oil exploitation.

The North Caucasus fields — North of the Caucasus, in the Russian Republic, oil has been produced for a long time at Groznyj and Majkop and these fields in the late 1960s approached Baku in their combined output. Much of the Soviet petroleum exports to Western Europe originate in the districts north and south of the Caucasus and are shipped from terminals on the Black Sea, e. g. Novorossisk, Tuapse, and Batumi.

The Volga-Urals-region — Another large oil region, a Second Baku, was developed before the Second World War between the middle Volga and the Urals, marked by the cities Kuibyshev, Ufa and Perm. The real expansion in this region came after the war. In the late 1960s the Volga-Urals district almost reached the position Baku used to have, but within a vastly expanded total production. Petroleum from the Volga-Urals district is refined at several cities in the area and along the Volga. Some is piped to Comecon countries of Central Europe, primarily East Germany and Czechoslovakia, and some is exported by way of Baltic terminals, primarily Ventspils. Finland gets two-thirds of its oil from the Soviet Union.

Northwestern Siberia — A new oil region in the Soviet Union is the Tyumen district of northwestern Siberia along the middle course of the Ob River and around the estuaries of the Ob and Yenisey Rivers, including the Yamal Peninsula, which began production in 1964. This remote area is rapidly being developed into one of the major petroleum districts in the Soviet Union with $^1/_3$ of the 1974 oil production. Primarily oil is found in the south (Ust-Balyk, Megion, Samotlor and others), and natural gas mainly in the north. A 1040-km pipeline (40 in.) has been built to the Omsk refinery where crude from this district is fed into the east-west pipeline system, which extends to East Germany in the west and to Irkutsk in the east and eventually may reach Nakhodka on the Pacific Coast. In the early 1970s negotiations were carried on with Japan about exports of oil from the Tyumen district.

Negotiations about a Japanese participation in the construction of a pipe line to Nakhodka were far advanced at the time of the oil crisis 1973, when the fourfold price explosion turned all cost estimates upside down. The Soviet Union was negotiating with Japanese and American interests about investments on the order of 7 billion dollars in five natural resource projects in Siberia and the Far East, of which the natural gas field in the Yakutsk region and the oil pipe line from the Tyumen district were to take the lion's share. The negotiations were complicated by the strained relations between the Soviet Union and China. America and Japan wanted good relations with both countries. Also China held out oil deliveries to Japan, which in the long run might considerably exceed those under discussion with the Russians. The stepped-up oil (and gold) revenues of the Soviet Union may have made the country less interested in obtaining long-term credits. At the same time, the steep increase in the Japanese oil bill should have made Japan less capable of extending credits.

⟶

Fig. 16 — Oil production and pipelines in the Soviet Union. Orientation map based primarily on Atlas Rasvitija Chozjaistva i Kultury SSSR, Moskva 1967. European part above (a), Asiatic below (b).

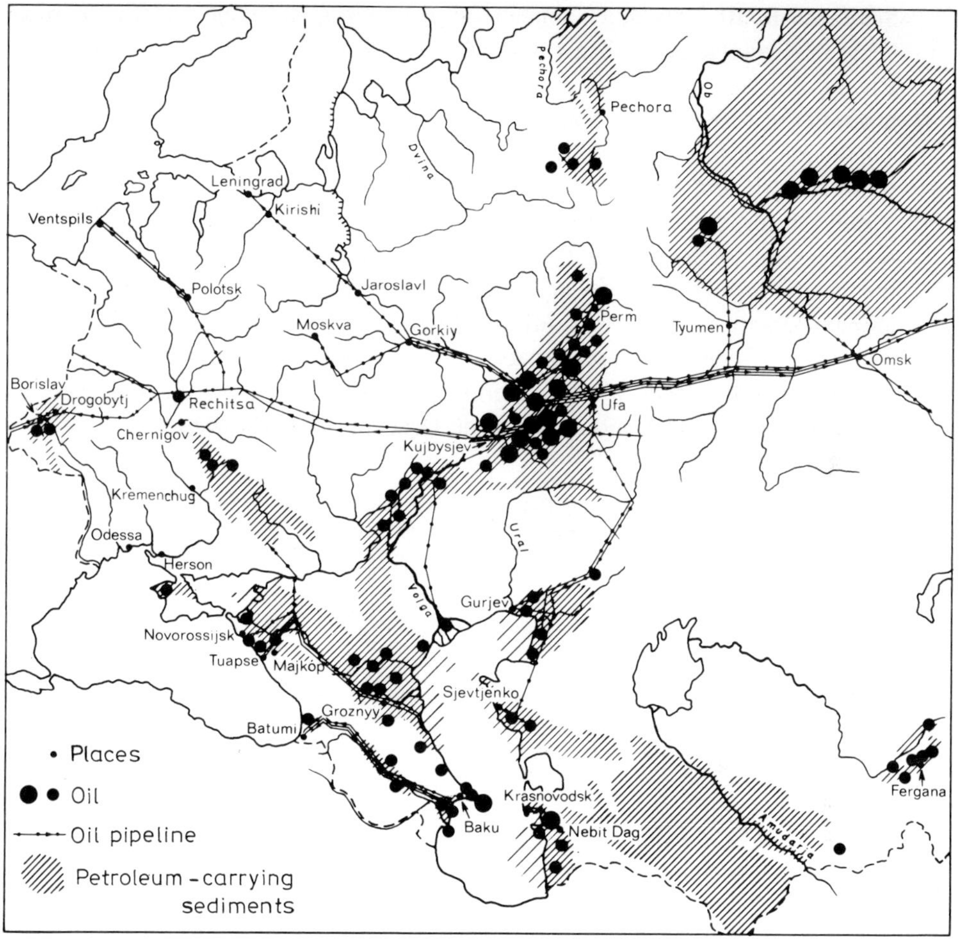

Pechora
Pechora
Ob
Dvina
Leningrad
Kirishi
Ventspils
Polotsk
Jaroslavl
Moskva
Gorkiy
Perm
Tyumen
Omsk
Borislav
Drogobytj
Rechitsa
Ufa
Chernigov
Kujbysjev
Kremenchug
Ural
Odessa
Herson
Volga
Gurjev
Novorossijsk
Tuapse
Majkop
Sjevtjenko
Groznyy
Batumi
Fergana
Krasnovodsk
Baku
Nebit Dag
Amudarja
Places
Oil
Oil pipeline
Petroleum-carrying sediments

a

Lena
Jenisei
Omsk
Novosibirsk
Irkutsk
Komsomolsk
Chabarovsk
Vladivostok
Nachodka

b

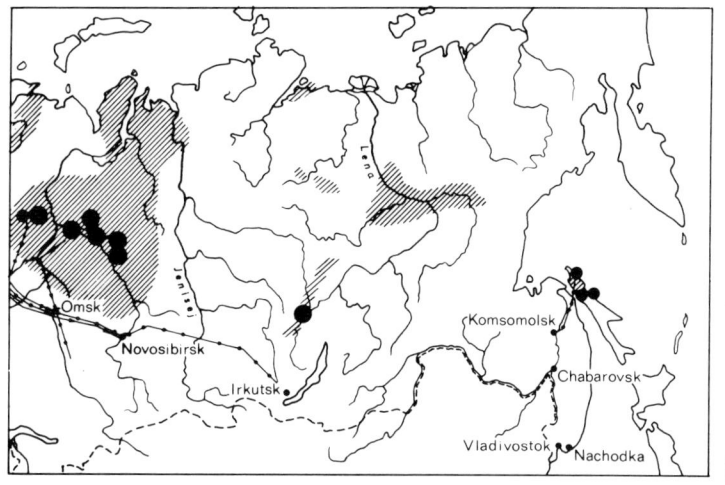

Scattered fields — A much smaller oil district was developed in the interwar years in Kazakhstan northeast of the Caspian Sea at the Lower Ural and Emba Rivers. Guryev is the chief center of this area. A promising oil discovery was made in the 1960s further south near Shevchenko on the Mangyshlak Peninsula. Turkmenistan has the Nebit-Dag oil district east of the Caspian Sea between Krasnovodsk and the Iranian border. The scattered deposits of the Fergana Valley straddle the border between Uzbekistan and Kirgizia. The Ukraine produces oil in the former Polish districts at Drogobych and Boryslav in the northern foothills of the Carpathians and in the Chernigov District of the Donets-Dnepr Upland. In southeast Belorussia oil was struck in the middle 1960s near Rechitsa; it appears to be a minor field. The Russian Republic, in addition to the fields already mentioned, produces oil in the Far East in northeastern Sakhalin and north of the Urals in the Pechora area.

Tab. 11 — Soviet Oil Production by Republic in Selected Years (per cent)

Republic	1913	1940	1950	1960	1972	1980
RSFSR	13	23	48	82	81	
Belorussia					1	
Azerbaijan	75	72	39	12	5	
Georgia		0.3	0.3	0	0.0	
Ukraine	10	1	0.6	1	4	
Kazakhstan		0.3	3	1	5	
Turkmenistan	1	2	5	4	4	
Uzbekistan	1	2	3	1	0.3	
Kirgizia		0	0.1	0.3	0.1	
	100	100	100	100	100	
Total, tons	10	31	38	148	400	700 (plan)

Source: Johannes Humlum, Kulturgeografisk Atlas, II Tekstbind, 7th ed. (Gyldendal: Aarhus 1971). Narodnoe Chosjaistvo v 1972 g.

Pipelines — The Soviet Union has not yet built out its pipeline system to match the increase in oil production. Early 1968 it was 32 200 km, just $1/_{10}$ of the American net (321 900 km). The railroads in 1966 transported almost 40% of the oil or almost six times as much as they handled in 1950. Serious efforts are made to move ever larger shares of the oil through successively wider pipelines. The east-west trunkline reached Omsk in 1955, Novosibirsk 1959, and Angarsk near Irkutsk 1964. Omsk is the major petroleum center of Siberia and its position is strengthened as production grows in the Tyumen district to the north.

The trunkline westward from the Volga-Urals district to central Europe, the Druzhba (Friendship) pipeline, was completed in the early 1960s. At Mozyr in

Belorussia it bifurcates into a northern line to refineries at Plock, Poland, and Schwedt, East Germany, and a southern line to Bratislava, Czechoslovakia, and Szazhalombatta, Hungary. A second trunkline westward from the Volga-Urals district carries oil to refineries at Gorky, Yaroslavl and Kirishi, southeast of Leningrad.

A 1600 km pipeline from Kuibyshev to Novorossisk, completed in 1974, permits exports from the Tyumen district of Siberia. Some 30% of the Soviet oil export (118 million tons) in 1973 went by way of this Black Sea port. The Soviet Union in the same year imported 15 million tons, primarily from Iraq and mainly for political reasons. A new terminal was opened in 1974 at Ventspils, which increases the capacity of the chief Baltic outlet from 15 to 20 million tons.

Prospects – The vast shelf areas north of the Soviet Union have not yet been prospected, but they should contain rich resources for the future.

The Soviet Union has large natural resources on-shore and is under no pressure to become pioneer in off-shore prospecting. The Soviets can calmly wait until the technique has been developed elsewhere, chiefly in the United States. They are, presumably, in less of a hurry than their neighbours to draw lines on the shelves for prospecting rights.

The Gulf

The tremendous oil acoumulations in the countries bordering the Gulf in the early 1970s accounted for 35% of world production. They dominated exports and held 60% of the known reserves. Through the 42 km wide Strait of Hormuz, named for a city of Persian antiquity, moved 1 billion tons of oil, almost half the consumption of countries with market economies.

This oil district was tapped already before World War I, when fields were developed in the western foothills of the Zagros Mountains of southwestern Iran, where oil and gas seeps had been known since Antiquity. Unsuccessful wildcats were drilled in 1884 and the first successful well went into operation in 1908 near Masjid-i-Sulaiman. The Anglo-Persian Oil Company, which succeeded Burmah Oil as an operator in 1914, had the British Government (the Admiralty) as the major stockholder. Road and pipeline were constructed to the newbuilt refinery at Abadan on the Shatt al Arab in 1913. During the Second World War, this refinery was to play an important part; at the end of the war it had been enlarged to a capacity of 25 million tons, the world's largest.

Iran – Production in the Iranian fields doubled between 1945 and 1951, when the Iranian Government cancelled the concession of the British company, renamed Anglo-Iranian Oil Company in 1935, and again changing name to

British Petroleum Company (BP) after its holdings in Iran had been expropriated.

For three years Iran produced oil only for its domestic market since exports without the consent of the British corporation was against international law. An agreement was reached in 1954 after American pressure. A consortium was formed by BP (40%), Shell (14%), Compagnie Française de Pétroles (6%) and 12 American companies (40%), which wrote a 25-year agreement with the National Iranian Oil Co about prospecting and production on contract with the state-owned company. Italian (Agip), American (Standard Oil Indiana), and Canadian (Sapphire) companies entered into joint ventures (50:50) with the National Iranian about development of lesser concession areas.

Access to Abadan is limited by sand bars in the Shatt al Arab; supertankers cannot enter the river. The refinery is listed with a lower capacity than before the 1951 oil conflict and the city of Abadan, with 270000 people entirely dependent on its refinery, may be seen as the first depressed area of the dynamic petroleum industry. New export terminals were built, first at Bandar Mashur in 1948 and in the early 1960s on Kharg Island, where tankers of more than 200000 tons can load.

With an oil production exceeding 250 million tons a year in the early 1970s, Iran was the second-largest oil country in the Middle East. Oil accounted for 90% of Iran's export value and oil revenue played a prominent part in the rapid economic development of the country.

Through the oil crisis of 1973/74 Iran's large oil revenues increased manyfold and gave the country unexpected possibilities of economic expansion. In contrast to other oil countries of the Middle East, Iran, with its 31 million people, has the population base for such development. In contrast to the Arab countries, Iran has refused to use oil as a geopolitical weapon. It sells oil to the multinational corporations and does not try to influence their sales of the commodity. Within OPEC, Iran, with relatively small reserves, is reported to be one of the strongest advocates for the high price.

Like some oil countries, Iran also has large reserves of natural gas. However, major markets are far away. A first step to develop exports of natural gas on a large scale was a triangle deal 1973 with West Germany and the Soviet Union. Gas from Iran will be fed into the Soviet pipeline system and the Soviets will deliver equal amounts to Germany from their European system.

Iraq – Oil concessions in Iraq date from the end of last century but the first well in the rich Kirkuk field north of Baghdad were not drilled until 1927. Large-scale production started in 1934 with the completion of two 30 cm (12-inch) pipelines to the Mediterranean ports of Haifa and Tripoli. After the Second World War, 40 cm (16-inch) lines were laid out parallel to the old ones, but, owing to the Arab-Israeli conflict, the second line to Haifa was never completed. In addition to French (Tripoli) and British-Dutch (Haifa) interests, American money was also involved in the development of Iraq's oil deposits.[12] The completion of the pipeline to Tripoli in 1934 coincided with the development of the domestic refining industry in France. The Shell refinery at Haifa, when cut off from its crude supply in Iraq, organized a trans-Israelian pipeline from Eilat on the Red Sea, which allows receipt of oil from Iran and other countries. A fifth pipeline from Kirkuk, the 75 cm (30-inch) line to Baniyas, Syria, was completed in 1952 followed by a sixth heavy line. The combined capacity of the pipelines to Baniyas and Tripoli is 60 million tons a year. A third terminal is built at Dortyol north of Iskenderun, Turkey. The 1040 km pipeline will have a capacity of 35 million tons a year.

Two oil fields southwest of Basra came into operation after WW II. Zubair (1949) and Rumaila (1953) are connected with Faw on the Shatt al Arab, which is limited to small tankers. The pipeline has been extended to a terminal 20 km offshore, Khor al Amaya, which takes tankers of 300 000 tons. Production in the two fields is expected to increase from 35 million tons in 1974 to 85 million tons 1978.

Iraq's oil production was long on a par with that of Iran, but an instable political situation and conflicts with the IPC has kept back the increase. Oil experts consider Iraq's reserves to be very large, surpassed only by those of Saudi Arabia. Both Kirkuk and Rumaila rank among the six largest oil fields in the world. Iraq's production may exceed that of Iran in the 1980s. For increased flexibility, the Kirkuk field will get a two-way pipeline to Faw so that oil from both regions may move either by way of the Gulf or the Mediterranean.

Bahrain – The first of the multitude of oilfields in and near the Persian Gulf were developed in the British-protected Sheikdom of Bahrain. After unsuccessful British explorations in the island, American interests took over and found oil in 1932. Bahrain was the first area where the American gained a foothold in the British-dominated Gulf Region. However, oil production in Bahrain has remained modest. The refinery and loading terminal of the island primarily serve as an outlet for Saudi Arabian oil.

Saudi Arabia – The American success in Bahrain made the neighbors in Saudi Arabia invite American companies to prospect for oil on the mainland. Standard Oil of California in 1933 formed the Delaware-registered Arabian American Oil Co (Aramco), which three other American oil companies later joined: Texaco, Exxon, and Mobil. In the early 1970s, Aramco produced more oil than any other company. But the starting-up period was long. Although oil was found in the 1930s, development of the vast deposits and production belong to the postwar period. Today, Saudi Arabia has the largest reserves in the world and the lowest production costs. The Damman field, producing since 1938, was soon followed by a much larger field at Abqaiq (1940), the Qatif field (1945) and after 1948, by the giant Ghawar field, one of the most extensive fields in the world. In 1951 oil was also found offshore at Safaniyah at the border between Saudi Arabia and the Neutral Zone.

Oil from Saudi Arabia has outlets in the Gulf as well as on the Mediterranean. Refineries and oil terminals are located at Ras Tanurah and Ras Khafji and in Bahrain Island. The longest pipeline in the Middle East, the Trans-Arabian Pipeline or the Tapline, was completed in 1950 and carries crude from the Abqaiq field to offshore loading facilities at Sidon, Lebanon. This 75 cm (30-inch) line has a capacity of 25 million tons a year.

In times of high freight rates for oil, the Tap-line was utilized to full capacity; when freight rates were at their bottom level, it was cheaper to fetch the oil with tankers in the Gulf, and capacity-utilization was only some 30–40%. In the beginning of 1975, when the bottom had fallen out of the tanker market and the Tap-line showed extra heavy losses, the pipeline was closed but later reopened for non-economic reasons.

Petroleum accounts for more than 95% of the revenue of the state and marks economic life in modern Saudi Arabia.[13] In the Hasa Province on the Gulf, where all oil fields are located, it finds many expressions in the landscape. Dharan, a company town, built by Aramco after 1938, is a typical American city with supermarkets, golf course, and movie theater, and with one of the busiest airports in the Middle East. Some 1 000 American employees and their families live in this city.

The explosion in government revenue from oil after 1973 was followed by elated economic planning. The refinery capacity in Saudi Arabia was to increase tenfold in three years from 1974, a new large steelworks was to be built in four years, and a large petrochemical complex in five years. Large irrigation projects are planned, both with dams to utilize the scarce rainfall and with drilled wells to utilize fossil water and groundwater originating in the mountains along the coast in the southwest. The feasibility of utilizing nuclear power for desalting sea-water in the Red Sea is being investigated and so is a project to take fresh water as a return cargo in the supertankers. Before the infrastructure has been built out, the people have been trained for life in an industrial society and technical skills have been accumulated, most of the oil money must be put to work abroad. Capital from countries with a large oil production and a small population (Saudi Arabia, Kuwait, United Arab Emirates, Libya) will for many years move around as international liquidity and create a new type of problems in the financial world.

Kuwait is surpassed only by Saudi Arabia and Iraq in oil reserves. Concession to look for oil was granted in 1934 to British and American interests, the present BP and Gulf. Burgan, now the world's largest oil field, was discovered in 1937, but the first shipments were not made until after the war (1946). The pipe lines focus on Mena al Ahmadi, which ranks with Kharg Island and Ras Tanurah, as the leading export terminals for oil in the world. The population of Kuwait doubled between 1963 and 1970 (to 757000), most of them foreign-born. Also the Neutral Zone between Kuwait and Saudi Arabia is an important oil producing area. The revenues from oil in this zone are divided between the two countries.

The United Arab Emirates, with its 200000 people, is even smaller than Kuwait in population. Oil has been produced from 1962 in Abu Dhabi and from 1969 in Dubai, the two largest of seven sheikdoms. Like neighboring Qatar, the federation is now an important oil producer.

Company – State relations – In 1972 an agreement was reached with the oil companies in Saudi Arabia and Kuwait that the state should take over 25% of the oil investments in respective country. This share was to be increased to 51% in 1982. But already in 1974 the share was changed to 60% and in 1975 to 100% i. e. oil production according to the Iranian model.

From 'sleeping partners' of the oil companies, satisfied with a share of the land rent for oil which after negotiations with the operator was increased from time to time at the expense of the oil company, the producing countries with their successful cartel have become monopolists, who for the time being dictate the oil prices in the world market. A price much higher than the marginal cost at the most expensive well should, according to economic theory, lead to over-supply of energy within a rather short time and falling prices.

The multinational oil corporations, pinched between the ambitions of the producing countries to own and manage their oil fields and of the consuming countries to control in detail the supply and distribution of oil, can be expected to look for new fields of activities. To be a middleman between two state-owned oil companies cannot be satisfactory to the managerial and technical expertise of a multinational oil corporation. The development of alternative forms of energy, primarily coal and nuclear power, has already tempted many oil corporations (the term 'energy company' is common), but also the petrochemical industry and petroleum prospecting in other parts of the world, primarily in inaccessible areas, e. g. tropical rain forests and arctic regions. The most important field of activity will be the prospecting for oil and development of oil fields in the continental shelves, which are expected to hold as much oil as the continents.

Oil and geopolitics – The use by the Arab countries in October, 1973, in a regional conflict (the Yom Kippur War) of the oil embargo against distant countries like the United States, the Netherlands, and Denmark came as a complete surprise, not only to ship owners and shipyards, but to politicians and the business community at large. The oil embargo and the four-fold price-increase, unilaterally enforced by the oil cartel, demonstrated that the days of free trade in raw materials had come to an end. The new energy policy of the industrial countries must depart from this fact, but it will take several years before their vulnerability has been lessened. For the rest of the 1970s and into the 1980s the threat of economic blackmailing will press heavily on the geopolitical freedom of choice of the industrial countries. The Conference on International Economic Cooperation meeting in Paris 1976 was a dialogue between rich and poor nations, a north-south dialogue. OPEC held the Third World countries in a common block against the industrial market economies. For further discussion of oil and geopolitics, see pages 207–217.

Africa

Until the 1950s the Egyptian fields at Sudr and other places near the Gulf of Suez were the only oil fields of any importance in Africa. Today Egypt is a small producer compared with Algeria, where oil was found in 1955, Nigeria (1938), and Libya (1961). Algeria and Libya give further strength to the role of the Arab countries in the petroleum exports of the world. Egypt may also become an oil exporter in the future as a result of recent prospecting.

Algeria – The interest became focused on the Sahara as a petroleum region when the French oil company reported important finds in Algeria. Long pipelines, some 700–800 km, connect the fields at Hassi Messaoud and Edjeleh with the Mediterranean ports, Arzew, Bejaia, and La Skhirra (in Tunisia). In addition to oil, Algeria has a rapidly expanding production of natural gas for the domestic market and for exports by LNG tankers (Arzew).

Libya – The large oil fields of Libya have undergone a rapid expansion. The first was opened in 1959 some 200 km south of the Mediterranean with pipeline to Marsa el Brega. It turned out to be a major field, Zelten. In a few years the desert country of Libya became one of the large oil exporting countries. Short distances to the markets in Europe and North America and a location on the 'market side' of the Suez Canal contributed to the exceptional expansion. New, large oil terminals were built at Es Sider, Ras Lanuf, Zueifina, and Marsa el Hariga. In contrast to older oil producing countries, Libya has had a large number of oil companies participate in the development. When OPEC from 1970 tried to raise oil prices by cutting back production, Libya had to carry the heaviest burden. The 1974 production was just half the 1970 output.

During the reign of the holy King Idris, the American oil company Occidental in 1968 started an irrigation project in Kufra, the religious center of the Sanusian sect. Water is pumped from a well (76 l/sec) and is spread with fertilizers through a 560 m irrigation arm that moves on a motor-powered wheel, one rotation per day. Around every well is a circular, intensily cultivated area surrounded by desert. The ground water table is sinking more slowly than anticipated, the fossil water is expected to last for centuries. The successful irrigation project was nationalized after the coup d'état in 1969. New irrigation projects are being planned, i. e. in Sarir where BP developed one of the largest oil fields in the world.

Nigeria – The oil fields are located in the Niger delta and on the shelf. Prospecting started in 1938 but the upheavals of the war and the large postwar investments of Shell and BP elsewhere delayed the development. Other companies made the first successful drillings in this difficult area, where climate, vegetation as well as topography combined to make life hard for the prospectors. But Nigeria's favorable location and, in the beginning, economically attractive conditions for the oil companies led to a rapid expansion, when oil had been found. However, the civil war almost led to a standstill in 1968 since the oil fields to a large extent are in Biafra. Bonny Islands, Forcados, and Okan are export terminals for the oil.

Other Africa – New oil fields have been developed in Angola, not the least in the Cabinda exclave north of the Zaire River which is said to be a new Kuwait, as well as in Gabon and Zaire. After the 'deluge' of 1973 prospecting started in Chad, Cameroon, Niger, Ghana and other countries. But also multinational oil corporations have limited resources (technical equipment and skilled personnel). Many areas remain to be thoroughly investigated in Africa. In 1970, Africa accounted for 290 million tons or 13% of world production.

Other Asia

The populous Monsoon Asia has much less than its share of the world's oil production but may be more favorably equipped with potential reserves. Prospecting with modern techniques, for natural reasons, has been more intense in areas closer to the large markets of the North Atlantic region. For decades, wars have cut off China and Southeast Asia from serious prospecting efforts.

The British and Dutch colonies in Asia early played an important role for the supply of oil to the empires. Already before the turn of the century, the fields in Burma, southern Sumatra, eastern Java and eastern and northeastern Borneo (now Kalimantan) were producing.

Indonesia – Present Indonesia was an important oil producer in the interwar years. After wartime destruction the Dutch restored the oil fields and the terminals for a production of some 10 million tons by 1948. But friction between the oil companies and the new Indonesian government prevented further expansion.

The coup d'état of 1966 made possible a return to earlier economic relations with the market economy countries. The rapidly expanding Japanese oil market, almost exclusively dependent on oil from the Gulf countries, looked for alternative sources of supply, preferably closer to home. In the beginning of the 1970s, Indonesia was — along with the North Sea — the preferred area for oil company investments. At the time of the October Crisis of 1973, the Indonesian production approached an annual level of 75 million tons and more was to be expected.

In addition to already proven reserves in the far-flung island realm of Indonesia, manyfold larger deposits can be expected in the wide continental shelf, stretching from the Indonesian islands in the south to Korea and Japan in the north, when the sophisticated prospecting technique of the oil companies is brought to bear on a large scale. Oil has already been found off Sarawak and Sabah in East Malaysia and many consortia, representing a large number of companies, in 1973 and 1974 had obtained concessions to prospect for oil in the South China Sea, in the Gulf of Siam and around the coast of South Vietnam, where the first test well 300 km south of the Saigon beach suburb of Vung Tau in August 1974 showed traces of oil. In contrast to the North Sea, the shelf of East Asia has not yet been divided by the coastal states. Large-scale prospecting has to await the settlement of this controversial and difficult issue.

China — The energy balance of China until the mid-1970s was completely dominated by coal (some 85%). The recent rapid expansion of oil production, with annual growth rates of more than 20%, went parallel with a slow increase in coal production. The first oil fields were those of Sinkiang — Karamai in the northwest near the Soviet border and Yumen — followed after 1959—60 by the large deposit at Taching west of Harbin in Heilungkiang.

Taching is presumed to account for almost half the Chinese production. A pipe line was completed in the beginning of 1974 to Qinhuangdao (Chingwangtao) on the Yellow Sea. Other fields have been developed near the coast, at Shengli in Shantung (1964) and in the coastal marshes south of Tientsin (Takang, 1967). With a production of 80 million tons in 1975, China was on a par with Canada and Iraq as an oil producer. Ten years earlier, China turned out 10 million tons.

Even if the annual growth rate slows down considerably, China may soon surpass Saudi Arabia as an oil producer. Her future position among the oil exporters is more difficult to judge. Exports to Japan of 50 million tons by 1980 have been mentioned. China needs the foreign currency provided by oil exports for her development. On the other hand, the emphasis on mechanization of agriculture in her development program leads to a large consumption of oil and so does the decentralization program. Large, centralized production units benefit from economies of scale and modern techniques and often have the option of using coal or oil as a fuel; small and scattered units are more dependent on easily transported fuels like oil or natural gas.

The largest Chinese petroleum reserves should be found in the shelves off China's coast. A geological survey analysis of the shelves off East Asia was made in the late 1960s under the auspices of the United Nations and indicated several interesting structures. The first Chinese drilling took place in Bohai in the Yellow Sea. Expertise on off-shore prospecting and drilling is strongly concentrated to American companies. With present competition for off-shore drilling platforms, technical expertise and personnel it may, however, take a long time before China becomes a large producer of off-shore oil. The geopolitical complexities of the region may also delay such production.

India has remained a small oil producer with fields in Assam and at the Gulf of Cambay. Large parts of the Indian subcontinent have a geological structure that precludes finds of oil while others, not the least the Bay of Bengal, have heavy packs of sediments. It has been insinuated that the oil companies have shown a minimum of interest in finding oil in India, preferring to sell cheaply produced oil from their fields in the Persian Gulf countries. In recent years the Soviet Union has assisted India with technique and personnel for oil prospecting.

Burma – From 1886 Burma was the base of operation of one of the oldest oil companies in the world, Burmah Oil, and an important exporter. 'Burma's way to socialism', introduced as a national goal after the coup d'état 1962, turned out to be economic isolationism. The oil industry was nationalized and production is now only for domestic consumption. Production is of the same magnitude as in Japan or Pakistan, less than a million tons a year. After a catastrophic decline in her rice exports and loss of her main source of foreign currency, Burma in 1975 departed from her principles and entered into negotiations with Japanese, American and European interests about off-shore oil prospecting.

Oceania

Australia – Petroleum was discovered in the stormy Bass Strait off Gippsland, Victoria, in 1967. Oil and gas are piped ashore at Westernport southeast of Melbourne, where a heavy industrial complex is being built, based on petroleum and steel and centered around one of the few natural deep harbors in the country. In the mid-1970s, Australia supplied about 70% of its own needs of oil, of great importance for the balance of payments in a country with one of the highest densities of cars in the world.

Prospects of the Oil Industry

The size of the remaining oil reserves has long been a controversial subject. The opinions of oil geologists vary. The discussion always concerns the amount of oil that may be obtained with present oil field practices. The share of recoverable oil varies from field to field but on the average should not much exceed 30% of the oil in the ground.

The heavy costs for prospecting and outlining oil fields prevent the oil companies like other mineral industries from looking more than a few decades ahead. Proven reserves are normally larger at the end of a year than at the beginning, in spite of production. With a continued high rate of increase of consumption, figure 15, oil reserves would relatively soon be exhausted irrespective of their size.

Against a continued rate of increase of more than 7% a year speaks the ever increasing share of oil in the energy balance. The total energy consumption in recent decades has had a significantly lower rate of increase (5%), figure 5. The continued high rate of increase for oil in the third quarter-century can partly be explained by the switch from coal to oil, a unique occurence. Against a continued high rate of increase speaks above all a high oil price. In the two years after 1973 neither oil nor total energy reported any increase for world production.

The confrontation between export countries and market countries is likely to lead the market countries to abolish their laissez-faire-policy in the field of energy. Cooperation among the industrial countries in developing new sources of energy and oil fields under their own control is to be expected. Much of the world's oil may then be left undiscovered or unused.

The hydrocarbons that tar sands and oil shales contain or that can be synthesized from coal represent very large quantities in comparison with direct petroleum reserves; how large depends on definitions. Each category represents more fuel than oil and gas combined, flowing and nonflowing. But new energy sources must be developed within some decades if the petrochemical industry of the future shall have its raw material base ensured.

Natural Gas

Natural gas in 1971 accounted for 18% of the energy consumption in the world and in 1980 is expected to reach 21%. Only three regions exceed or reach the average: United States (35%), Canada (26%) and the Soviet Union (18%). All three are countries, which means a unified political control of the long distant pipe-line-system. Where the production area is separated from the market by one or more frontiers a close economic cooperation between two or more

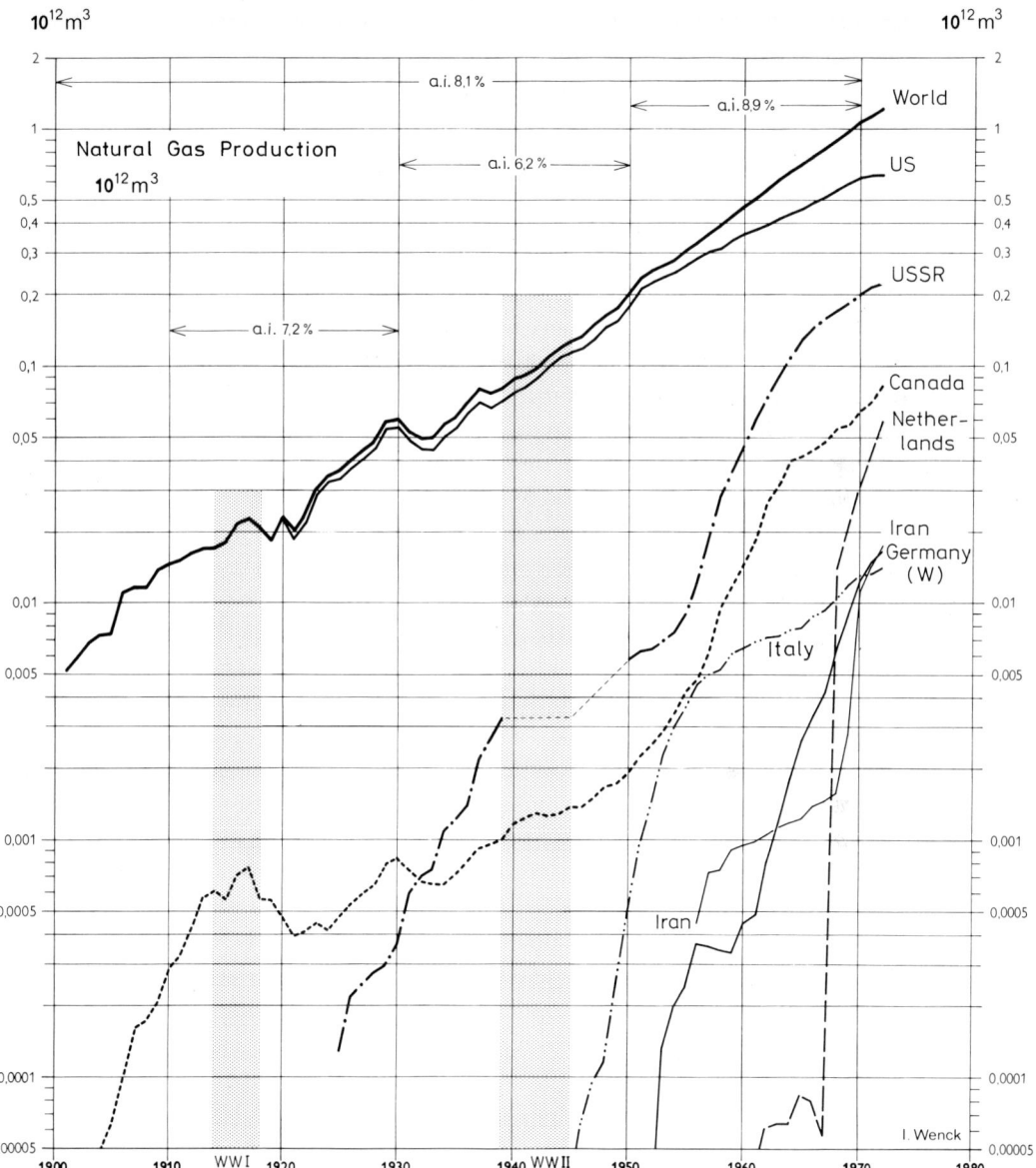

Fig. 17 – After the Second World War, natural gas has had a more rapid expansion than oil. It is a pure fuel, competitive with oil and coal in spite of higher prices. Still in 1974, the gas flared off in oil fields was estimated at 370 million tons of oil equivalents. The rapid increase in the fleet of LPG and LNG tankers makes it possible to export gas also from peripheral fields. A few industrial countries, among them Sweden, still do not have access to natural gas.

nations is a prerequisite. Such cooperation is being established in several parts of the world, even between countries with different economic systems.

Natural gas is more difficult to transport than oil because of its lower heat value per unit of volume. Expressed in tons of oil equivalents, natural gas corresponds to 0.00079 ton per cubic meter while oil amounts to some 0.86 ton. From the point of view of the household, natural gas is easier to handle than oil, which is more convenient than coal. Natural gas is almost like electricity. In the United States oil took the place of coal (and anthracite) as a household fuel from the 1920s to be replaced in its turn after World War II by natural gas. For the sake of convenience the consumer is willing to pay more for a given heat amount of oil than for coal and more for natural gas than for oil. Natural gas is a clean fuel, an advantage that becomes more obvious with every new restriction against sulphur effluvia in the air.

Natural gas has always been found in small amounts in the form of marsh gas (methane, CH_4) which is formed by vegetable decay in marshy ground. As mine gas it has caused many accidents since mixed with air it is highly explosive. In large quantities natural gas became available with the growth of the petroleum industry. During the pioneer period there was a market only for kerosene, gasolene and lubricant; the gas and the heavy fractions went to waste. Still gas is flared off in peripheral oil fields. In 1974 the amount was estimated at 350 million coal equivalent tons. Much gas is pressed back into the reservoir to keep the pressure and make possible a larger recovery of oil.

In oil and gas fields are found not only natural gas (primarily methane) but also petroleum gas (primarily butane and propane). The latter in many uses is an alternative to natural gas. It is known commercially as liquefied petroleum gas (LPG). It can relatively easily be transformed into liquid form and can therefore be transported in rather simple tankers. LPG is obtained at oil and gas fields but also at refineries. Before the crude can be shipped it must be stabilized, i. e. the surplus propane and some butane must be removed to lower the gas pressure. Increasingly this gas is being condensed and compressed for shipment by pipe line or tanker to a market. Until recently it was flared off. The gas flames were typical elements of the landscape in oil fields and at loading terminals.

Natural gas can also be compressed to fluid form for transport over the sea, liquefied natural gas (LNG). It is a rather expensive fuel and not competitive in all markets. LNG requires heavy investments at the shipping and receiving terminals and in special tankers. Some 600 m^3 natural gas is compressed to 1 m^3 LNG which is transported in fluid form and then decompressed for distribution by way of a conventional pipeline system.

North America

United States – The occurrence of natural gas in the first oil fields of western Pennsylvania and adjacent parts of the Ohio river basin were of great importance for the early location of the energy intensive glass and aluminium industries. For a short time gas accounted for larger quantities of energy than oil, figure 6. But the important gas fields, some 90% of total reserves, are in Texas, Louisiana, Oklahoma, Kansas and New Mexico, far from the dominant energy markets of the Manufacturing Belt in the Northeast. It took a long time to build the pipe line system which now enmeshes the country. The system grew from 123000 km in 1945 to 407000 km in 1970. The sales of natural gas expanded from $110.9 \cdot 10^9$ m³ to $619.6 \cdot 10^9$ m³ in the same time. The equivalent amount of coal in 1970 would have been 795 million tons and of oil 530 million tons. The American production of natural gas in 1970 thus exceeded both the production of coal and of oil.

In 1945 less than half the produced gas was sold outside the state of production but in 1970 the share had reached ²/₃. Near the gas field natural gas is a cheap fuel. In the producing states 90% of the gas goes to electric power plants, oil fields and manufacturing industry while in the northeastern states the households account for 70% of the market. For power plants and manufacturing industry in the Northeast coal and oil are cheaper fuels than natural gas which has to carry heavy transfer costs. Households are willing to pay a higher price for a more convenient fuel.

In the early 1970s there was serious concern that the domestic natural gas reserves would soon be exhausted. The quotient of reserves through annual consumption which in 1950 was 26 : 1 and 1966 was 16 : 1 had declined to 11 : 1. United States in 1971 imported natural gas corresponding to 50 million tons of

Tab. 12 – Per Capita Consumption of Natural Gas and Energy in the United States

Year	Natural gas 10^9 kWh	All energy 10^9 kWh	Natural gas in per cent of all energy
1920	2.3	54.5	4
1930	4.7	53.0	9
1940	6.2	53.0	12
1950	12.0	66.2	18
1960	20.8	73.0	28
1970	32.2*	99.0*	33

* Cumulative annual rate of increase between 1920 and 1970: natural gas 5.4% and all energy 1.2%. The same for the 1960s; natural gas 4.5% and all energy 3.1%.

Source: P. F. Corbett, "National Gas – Growth of a World Market" in K. A. D. Inglis, ed., Energy: From Surplus to Scarcity? (London: Applied Science Publishers, 1974), p. 48.

oil of which more than half was pipelined from Canada and the rest was LNG from transocean areas, primarily North Africa. The seaborne imports were expected to increase in the 1970s in spite of twice the domestic prices. The share of natural gas in the energy balance was expected to drop to 30% by 1980. It was against this background of diminishing domestic reserves that the American interests in the early 1970s in large-scale natural gas deals with the Soviet Union should be seen.

Even greater opportunities for the future is offered by methods of making synthetic natural gas (SNG) from other fossil fuels: coal or the petroleum fraction naphta. The principle of the process is simple: coal or naphta are being synthezised with the hydrogen of water at a high temperature and form methane. In practice the problem has turned out to be difficult and expensive. German industry has worked with it since 1910 (Bergius, Fischer-Tropsch). The gasification of naphta is simplest and many plants have recently been built or are under construction. They are expected to account for 15% of the gas consumption in 1985. But natural gas from a naphta basis runs into the same balance of payment problem as the oil imports.

A more long-term solution would be provided by the coal-based method since the United States has tremendous reserves of low-grade coal or coal with a high sulphur content or with both handicaps. A transformation of this energy to SNG or 'syn-gas' and distribution by the pipe line system for natural gas would be a way of utilizing the coal. According to plans in the mid-1970s for making America more or less independent of energy imports some 340 million tons of coal would be utilized for gasification by 1985. Another way of using the coal would be to make power gas for combined-cycle systems of power plants located

Tab. 13 – Costs in the United States for Some Clean Fuels (January 1972)*

Fuel	Mill** per kWh
Natural gas: Texas	0.55
Louisiana	0.68
Alaska	0.82
US, average at source	0.61
New York City	1.19–1.54
LNG, East Coast	2.73–3.41
Low-sulphur fuel oil, East Coast	2.22–2.56
SNG, naphta	3.75–4.10
SNG, coal-Lurgi	3.58–3.92
SNG, coal – American methods***	2.39–3.24

* Notice that data refer to l'ancien regime in the energy field.
** 1 mill = 0.001 dollar.
*** Assumes opencast mining in the West.
Source: A. L. Hammond, W. D. Metz & T. H. Maugh II, Energy and the Future.

at the coal field. High-voltage transmission lines would then transport the energy to the market.

As seen from table 14 American energy projections from the early 1970s assumed a steeply decreasing increase in energy demand during the rest of the century. In absolute amounts all energy forms were to increase and the largest growth would be registered for nuclear power.

Tab. 14 – Projection of Energy Demand in the United States 1970–2000

Energy form	1970	1975	1985	2000	Annual increase rate 1970–2000, %
Oil	43.1	40.8	35.6	34.6	2.7
Natural gas	32.8	32.4	29.5	26.4	2.7
Coal	20.0	18.2	16.7	13.7	2.2
Water power	3.8	3.2	2.6	2.6	2.1
Nuclear power	0.3	5.4	15.6	22.7	19.9
	100.0	100.0	100.0	100.0	
Joules · 10^8	72.9	93.9	141.5	203.0	
Annual increase, %		5.2	4.2	2.4	

Source: A. R. Ubbelohde, Paper in Revue Soc. Roy. Belge des Ingenieurs et des Industriels, p. 229.

Canada – Some 40% of the Canadian natural gas is pipe-lined across the border to the United States, where Canadian gas makes up less than 5% of the consumption. The largest gas fields are associated with the oil fields in the prairie provinces. Pipe lines on Canadian territory supply the metropolitan cities. The gas reserves in the Northwest Territories, in the MacKenzie Delta and the Beaufort Sea, and in the adjacent American Prudhoe Bay may become available if the feasibility study undertaken by a 26-company consortium for construction of a 4 160 km pipe line (122 cm, 48 inches) arrives at a positive conclusion.

Western Europe

Until the middle 1960s natural gas was consumed relatively close to the gas fields and accounted for small but important shares of the energy balance in Italy, France, Germany and Austria. In 1960 natural gas supplied less than 2% of the total energy in the EEC. The discovery of the Groningen field in the northern Netherlands (Slochteren, 1959), one of the largest in the world, and several fields in the southern part of the North Sea has radically changed the

Tab. 15 – Consumption of Primary Energy in the EEC-6

Source of Energy	Per cent 1967	1972	Annual increase, %
Coal	30.3	17.5	– 4.7
Brown coal	4.7	3.6	0.9
Natural gas	4.6	12.3	29.5
Petroleum	53.9	61.5	9.5
Water power	6.3	4.9	1.1
Other sources	0.2	0.2	9.1
	100.0	100.0	6.5
Mt, coal equivalents	667.3	915.3	

Source: Verein Deutscher Kohlenimporteure, Jahresbericht 1973, p. 81.

situation. Natural gas in the late 1960s was the most rapidly expanding source of energy in the EEC-6.

More than 40% of the Dutch gas remains in the Netherlands, 25% is exported to West Germany, 13% to Belgium, 10% to France and 7% to Italy. The Soviet Union pipelines gas to several Comecon-countries and to Austria. Advanced plans exist for extending the Soviet pipe line system to northern Italy, southern Germany, France, and to the Ruhr-area. Gas from the Norwegian Ekofisk-field joins the West-European pipe line system at Emden in northern Germany. An important base area for supplying western Europe with natural gas is North Africa, primarily Algeria but also Libya (Marsa el Brega). A loading terminal for LNG from Algeria was built in the early 1960s on the Thames estuary from

Tab. 16 – Gas Consumption in Some European Countries 1971 and 1980 (Proj.)

Country	Mtoe Natural gas 1971	1980	Total energy 1971	1980	Natural gas in per cent of total energy 1971	1980
Germany (W)	16.8	52	235	340	7	15
Netherlands	21.1	40	61	100	35	40
Belgium	5.1	10	46	70	11	14
France	10.7	23	152	255	7	9
Italy	10.1	23	119	235	8	10
United Kingdom	18.5	50	206	275	9	18
Other countries	4.4	27	236	425	2	6

Mtoe = million ton oil equivalents.

Source: P. F. Corbett, "Natural Gas – Growth of a World Market" in K. A. D. Inglis, ed., Energy: From Surplus to Scarcity? (Barking: Appl. Sci. Publ., 1974).

where pipe lines were laid to several cities in England. Natural gas with its high heat value is cheaper than the earlier used city gas. The large gas finds in the southern part of the North Sea have radically changed the British energy situation. The gas from these fields is handled by the state-owned British Gas Corporation which also receives gas from the petroleum fields off the coast of Scotland which often contain both oil and gas.

Soviet Union

Before 1955 the Soviet oil industry was underdeveloped; natural gas at this time was an almost undeveloped natural resource. The following twenty years led to fundamental changes. The gas fields account for almost the entire production; only small amounts are obtained in oil fields and at refineries.

Prospecting in the vast Soviet territory led to discoveries of gas fields in widely scattered areas: at Saratov near the Volga-Ural oil district, at Krasnodar and Stavropol north of the Caucasus and at Sjebelinka in the eastern and at Dasjava in the western Ukraine. Large reserves were found beneath the Kara Kum and Kyzyl Kum deserts in Central Asia, at Gazli in Uzbekistan as well as to the left of the Amu-Darja in Turkmenistan. Later large gas fields were discovered west and north of the oil fields of the Tyumen district of northwestern Siberia as well as in eastern Siberia. The latest in the long line of large gas fields is Orenburg south of the Urals.[14] This field is expected to become the raw material base of a large petrochemical complex near Togliatti on the Volga, which may be built with the assistance of American interests.[15]

The Soviet pipe line system for gas was extended from 2 300 km in 1950 to 79 100 km at the end of 1972. The expansion primarily took place after 1958 when the net was 9 500 km. Since the net is new with long lines the Soviet Union as a rule has pipes of wider dimensions than the United States. Two 1 900 km lines (100 cm) connect Gazli and other fields in Central Asia with Tjeljabinsk, Sverdlovsk and other manufacturing centers in the Urals. The first was ready in 1963. The Urals also receive gas from the north, from West Siberian fields. In 1967 a 2 600 km line (100 cm) was opened between Central Asia and the Moscow region.

For the development of the most recent fields the pipe diameter was increased to 120 and 140 cm. The latter rapidly became standard for long lines. Heavier pipe allows less material and lower investment cost per transported volume gas. The heaviest pipe is used in lines from Tarko-Sale and Novij Port and other gas fields in the Tyumen district which by way of Vorkuta north of the Urals run to the Moscow area and Leningrad. The 'Northern Lights' continues to the bordertown Uzgorod with branches to Czechoslovakia, Poland, DDR, Hungary and so on.

Leningrad and Tallinn also receive gas from the exceptionally rich oil shale deposits at Kohtla-Jarve which yield up to 320 liters of oil to the ton.

Stockholm
Helsingfors
Tallin
Leningrad
Kochtla-Jarve
Pechora
Dvina
Novyy Port
Vorkuta
Pechora
Ob
Tarko Sale
Moskva
Tyumen
Chelyabinsk
Uzgorod
Dashava
Togliatti
Orenburg
Saratov
Ural
Volga
Shebelinka
Krasnodar
Stavropol
Gazli
Baku
Places
Oil
Gas-oil
Oil-gas
Gas-pipeline
Petroleum-carrying sediments

a

Vorkuta
Novyy Port
Pechora
Tarko Sale
Jenisei
Lena
Jakutsk

b

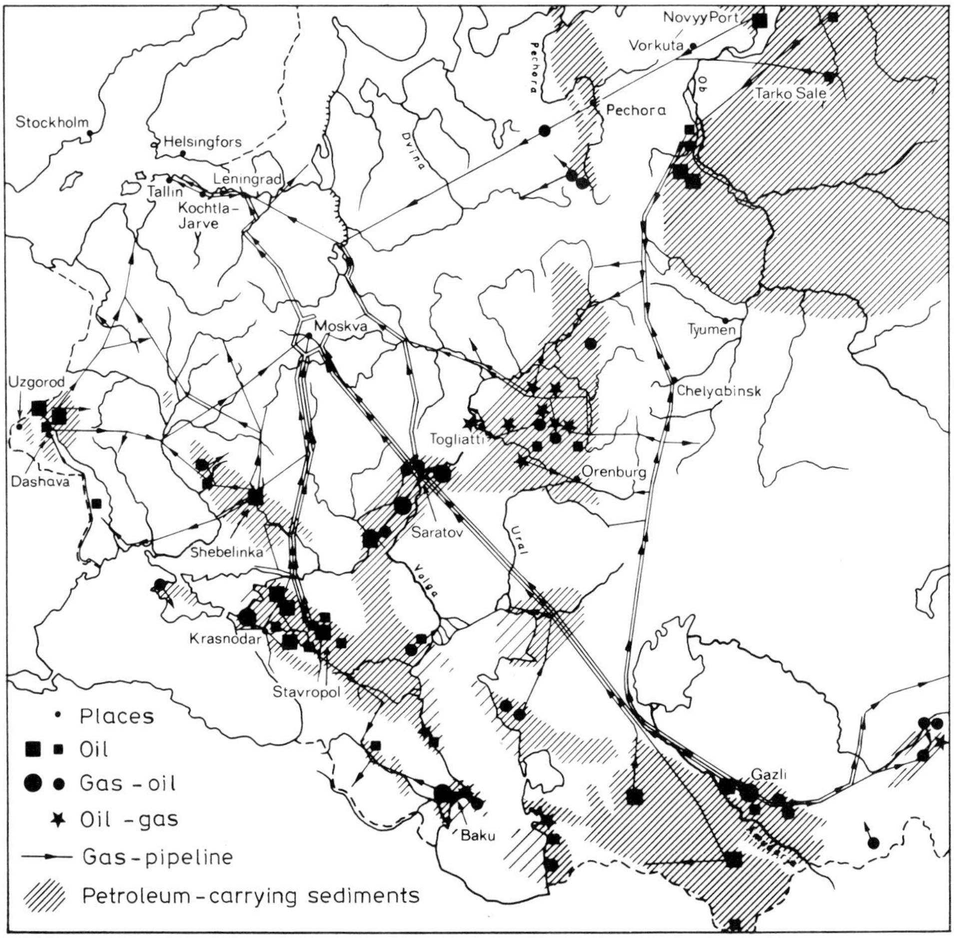

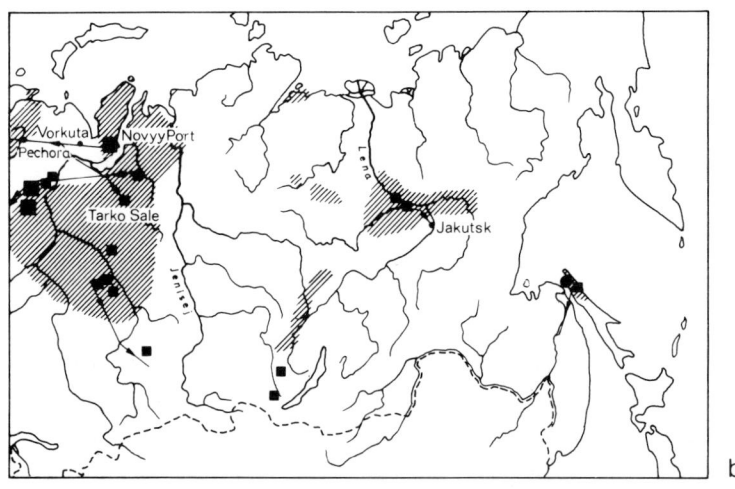

Potentially, oil shales may represent an even larger resource of fossil fuels than coal. In comparison with petroleum and nuclear power, oil shale and coal require extensive mining operations which would cause formidable environmental problems if they were to play a major role as energy carriers. The Estonian oil shales are of great theoretical interest, although they play a subordinate part in the Soviet energy balance. The cost of extracting hydrocarbons from most oil shales is still far too high. It has been calculated that 40 liters of oil would be used in processing one ton of shale (mining and distillation) which sets the minimum limit for a potential oil shale resource. However, some shales are rich in uranium (e. g. the Västergötland shales of Sweden) which means that they may be mined for their nuclear power with hydrocarbons as a byproduct.

Tab. 17 — Natural Gas Production in the Soviet Union Selected Years by Republic (in per cent)

Republic	1940	1950	1960	1965	1970	1972	1980
RSFSR	7		54	50	42	40	
Ukraine	15		32	31	31	30	
Belorussia	—		—	—	0	0	
Uzbekiztan	0		1	13	16	15	
Kazachstan	0		0	0	1	2	
Azerbaijan	78		13	5	3	3	
Kirgizia	—		0	0	0	0	
Tadzhikistan	0		—	0	0	0	
Turkmenistan	0		0	1	7	10	
	100		100	100	100	100	
Total, 10^9 m gas	3.2	5.8	45.3	127.7	197.9	221.4	700 (plan)
Mtoe	2.5	4.6	35.9	101.1	156.8	175.4	555 (plan)
Acc., aipc*		6.1	22.8	23.0	9.2	5.8	13.5**

* Accumulated, annual increase, per cent.
** Planned increase 1970–80.
Source: Narodnoe Chosjaistvo SSSR v 1972 g.

Fig. 18 — Natural gas production and pipelines in the Soviet Union. European part above (a). Asiatic below (b). For source, see fig. 16.

Nuclear Power

The atom bomb dropped from an American bomber on August 6, 1945, on the Japanese city of Hiroshima marks a turning point in the history of military policy. The nuclear armament race between the United States and the Soviet Union after the war created a new balance system in the world, the so-called balance of terror.

First the atom bomb, and then the hydrogen bomb in the early 1950s, made the general public aware of the tremendous amounts of energy bound in the nuclei of atoms. It was obvious that this energy could also be used for peaceful purposes. A period of exaggerated ideas about the rapid break-through of nuclear power followed. Many started to talk about the Atomic Age.

A reaction came during the period of the nuclear armament race with the discussion about the radioactive fallout and its genetic consequences. The drama of the atom bomb and the long term influence on the genes of radioactivity made the nuclear power discussion more emotionally loaded than any other.

The development of the nuclear technology ran into greater difficulties than could be predicted from the outset. The material development in the postwar period has been formed more by the inconspicuous transistor, undramatically presented to mankind in 1948, than by atomic energy. It is more realistic to talk about the Electronic Age than the Atomic Age.

The uranium bomb on Hiroshima was the result of an imaginary race during the war about which side was to be first with the atom bomb. After the war it turned out that Hitler early had come to the conclusion that the atom weapon would not be ready before the end of the war. German research, on a limited scale, was therefore directed to the use of atomic power as a source of energy.

In the United States thousands of scholars were mobilized, among them leading Europeans such as Enrico Fermi and Niels Bohr, and sent with secret orders to isolated research centers, e. g. Los Alamos near Santa Fe in New Mexico. Hitler's timetable turned out to be correct. The first atom bomb, tested at Alamogordo in the desert of New Mexico on July 16, 1945, was completed when the war in Europe was already over and Truman, Stalin and Churchill had met in Berlin. The second bomb was dropped on Hiroshima. The third, the plutonium bomb on Nagasaki, was the last one in the American arsenal.

History – Basic research that eventually led to the atom bomb can be traced back to the discovery in 1895 by the Würzburger professor Wilhelm Röntgen of mystic rays, by him named X-strahlen, which came from the cathode rays in a vacuum tube and led to fluorescense on a nearby platinum plate. Henri Becquerel of Paris in 1896 studied the relation between X-rays and fluorescense and found that uranium emits rays without connection with daylight. He had discovered radioactivity or the spontaneous disintegration of the atom nucleus and the emission of electromagnetic radiation. The couple Marie and Pierre Curie in 1898 discovered two more radioactive elements, named polonium and radium, when they found that the uranium mineral pitchblende yielded more radioactivity than corresponded to that of uranium.

Even the understanding of the arrangement of the atom was changed at this time. It was no longer seen as a cue ball but as something with an internal structure. The main features of this structure were clarified in the three decades before 1920. The most important contributions were made by two Cambridge professors, J. J. Thomson and Ernest Rutherford. The latter showed that one type of emission from radioactive material, for want of better called alfa rays, were helium atoms that had been deprived of their two electrons. By analyzing the deviation of the positive alfa rays when they passed through thin gold foil Rutherford drew conclusions about the minute size of the atom nucleus compared with the atom. He showed that radiation led to chemical changes in the atom, that one element could be changed into another, which old time alchemists had tried but never succeeded in doing.

The atom – The atom is made up of a positive nucleus surrounded by negatively charged electrons. The nucleus in its turn is made up of positive protons and neutral neutrons which collectively are known as nucleons. Only elements with the same number of electrons and protons exist in nature. A nucleon has 1840 times the mass of an electron which means that the mass of the atom for all practical purposes is the mass of the nucleus. This occupies a sphere with a diameter of no more than 10^{-12} cm or $1/10000$ of the diameter of the atom. The atom is almost entirely made up of a void. The electron has a mass of $9.107 \cdot 10^{-28}$ g and charge of $4.802 \cdot 10^{-10}$ statcoulomb. It moves around the nucleus and around its own axis with a magnetic moment of $1/2$ quantum unit. The electron also has wave characteristics, with a frequency and a wavelength.

The mass of the proton is $1.673 \cdot 10^{-24}$ g and of the neutron $1.675 \cdot 10^{-24}$ g. The number of nucleons is given by the mass number written to the left of the chemical symbol. Sometimes the atomic number is also indicated, but this is redundant since it is shown by the chemical symbol itself. An element is defined by the number of protons in its nucleus. This establishes its chemical identity in the periodic system (e. g. uranium 92) but the number of neutrons may vary. Thus uranium has several isotopes with the same atomic number but different mass number: ^{233}U, ^{234}U, ^{235}U etc. From a chemical point of view different isotopes of the same element are identical.

The mass of the atomic nucleus is always less than the sum of the mass of the nucleons. The mass deficiency of the nucleus is a measure of the mass that was changed into energy when the nucleus was formed.[16] The mass deficiency per nucleon varies from element to element and is a measure of the stability of the nucleus. High stability is associated with a low energy level (large mass deficiency). When elements are ranked according to their mass numbers, starting from the lowest, the mass deficiencies rapidly increase and reach a plateau among the medium heavy nuclei after which they again decline. The highest energy levels are therefore found among the lightest nuclei. But also the heaviest nuclei have a high energy level. This may sound paradoxical against the background of Einstein's formula. It should be remembered, however, that hardly anybody even dreams about changing an element into energy; only the small fraction represented by the mass deficiency is concerned.

In the lightest elements energy is released when the atom nuclei are merged or fused to heavier nuclei (greater stability and mass deficiency, lower energy level). In the heavy elements the same thing is achieved by a cleavage or fission of the nuclei.

If the hydrogen fusion in the sun could be copied on earth under human control sea water would provide fuel for human needs for millions of years. The United States detonated a hydrogen bomb already in 1952 and the Soviet Union in 1953 but the fusion of the heavy isotopes deuterium and tritium which form helium is uncontrolled in the hydrogen bomb. An uranium bomb is used as an 'igniter' to initiate the high temperature needed for a chain reaction. The hydrogen bomb left a radioactive fallout because of its igniter but the hydrogen power of the future will be clean.

Fission Energy

Uranium (and potentially also thorium) are the raw materials of fission energy. The Manhattan Project, initiated at Columbia University, culminated when Enrico Fermi on December 2, 1942 started the world's first nuclear reactor in the squash hall under the stands of the Stagg Field at the University of Chicago. The reactor had its origin in the discovery by O. Hahn, F. Strassmann, and Lise Meitner that a slow neutron could be made to cleave a nucleus of ^{235}U into one barium and one crypton nucleus. Fermi shut off the reactor after 28 minutes and thereby showed that the nuclear chain reaction was under control. Uranium-235 is the only natural isotope that is fissionable.

The reserves of high-grade uranium are still incompletely known. The most important minerals are pitchblende with 75–90% uranium oxide and carnotite (62–65%). With higher uranium prices the raw material base widens. A doubling of the price more than doubles the reserves and new countries enter the list. At a doubled price Sweden would take the 4th place in the Western World with 310 thousand tons (Ranstad, Västergötland).

For the economy of a conventional thermonuclear reactor the uranium price is of relatively small importance. For a breeder reactor the price of uranium will

Tab. 18 — Uranium oxide (U_3O_8; reserves 1970 and production capacity 1975) (thousand tons)

Country	Reserves: US $ 22 per kg	Production capacity per year
United States	227	21
Canada	210	12
South Africa	181	5.4
France	41	2.1
Niger	24	1.7
Australia*	20	1.4
Gabon	13	0.7
World**	761	46

* Should be revised upwards after large discoveries in 1970.
** The Communist countries not included. The pitchblende mine at Joachimstal in Czeckoslovakia, known since the beginning of the 16th century, delivered the ore from which radioactivity was discovered.

Source: SY 1972 and MY 1970.

play a very subordinate role. Even low-grade uranium ore will then be of interest.

Light Water Reactors

Nuclear fission as a source of commercial energy had a long initial phase with many obstacles before the first units could be built. The pioneer plant (5 000 kW) was opened at Obninsk southwest of Moscow in 1954.

The British nuclear power expert, Professor D. C. Leslie, made the following estimate of the costs of British electricity from nuclear power plants, with the corresponding data for thermoelectric plants fuelled with oil for comparison.

The nuclear power station in the table is assumed to cost US $279 and 464 a kilowatt. Construction time is six years, the interest rate 10% and costs for site,

Tab. 19 — Cost of Electricity from Nuclear Power Plant and From Oil-fuelled Power Plant

	Oil, total	Nuclear power, costs		
		fixed	variable	total
Low	0.367	0.297	0.10	0.397
High	0.614	0.494	0.10	0.594

Costs expressed in p/uso (uso = unit sent out from station).

road connections and power lines normal. With these costs added, total costs will be between US $355 and 593 a kilowatt.

With amortization in 25 years, 10% interest and 65% operational time the capital costs will be those shown in the table. The fuel cost is made up of three almost equal parts: the cost for natural uranium, the cost for enrichment and the cost for the manufacture of the fuel cells.

The price for electricity generated by nuclear power is sensitive to the interest rate of long-term capital but relatively insensitive to the price of uranium. Electricity from stations burning fossil fuels on the other hand will react with

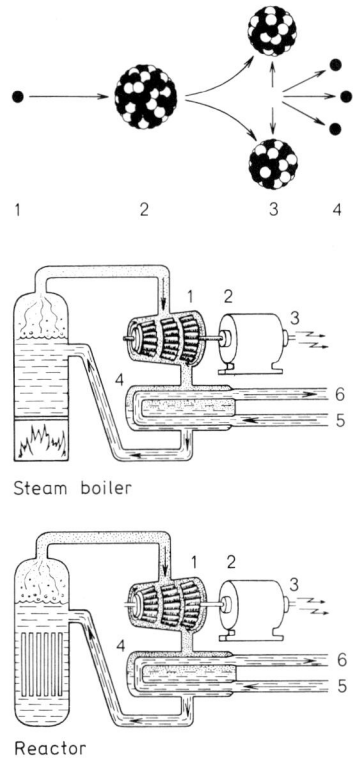

Fig. 19 – The functioning of a nuclear power plant
Nuclear fission – A slow neutron (1) hits a nucleus of uranium-235 (2), forms uranium 236, which is unstable and within a millionth of a second splits into two medium heavy nuclei (3), barium (Ba) and krypton (Kr), releasing energy, two-three fast neutrons (4) and gamma radiation. The probability of a chain reaction increases if the fast neutrons formed at the fission are slowed down. The neutron brake or moderator varies with the type of reactor.

A conventional thermoelectric power plant – The only difference between a conventional thermoelectric power plant and a nuclear power plant is the source of energy. Steam from the steam boiler (or reactor) powers the turbine (1) which in a generator (2) produces electric current (3) which by way of a high-voltage transmission line is fed into the power grid. From the turbine, steam moves to the condensator (4) where cooling water from e. g. the sea (5) in pipes and coils lowers the temperature of the steam below the condensation point. The heated sea water with its 'waste heat' is returned to the sea (6).

In an alternative plant, steam is not condensed but used for heating buildings, dehydrating brown coal etc. Such plants, built according to the counter-pressure principle, have a higher rate of efficiency. They are built on brown coal fields, or on the periphery of metropolitan cities.

Water in the reactor is highly purified and has to be reused over and over again. The steam boiler can be heated by coal, oil, natural gas, brown coal, wood, peat, straw, saw dust, household waste etc. but only the first four are used on a large scale. The reactor functions like a steam boiler.

The nuclear power plant – The reactor is contained in a room of concrete reinforced by steel-plate. The reactor hearth is made up of tens of thousands of vertical metal tubes with uranium dioxide. Water is boiled to steam and also serves as moderator (light water reactor). Two American light water reactors dominate the world market, the BWR (boiling water reactor), developed by General Electric, and the PWR (pressurized water reactor), launched by the other large electric machinery group, Westinghouse.

strong price fluctuations to changes in fuel prices but not be much influenced by shifts in the interest rate. A tenfold increase in the uranium price (US $18 a kilogram for U_3O_8 in 1973) would increase the low nuclear power price to 0.7 p/uso. A tenfold increase of the oil price from its summer 1973 level ($3 a barrel) increases the low electricity price to 2.0 p/uso. Seen in a different way the construction and engineering industries will benefit more than the mining industry from a nuclear power construction program.

Enrichment plants – Natural uranium contains only 0.71% of the fissile isotope U-235. Only the Canadian CANDU of the commercial reactors uses natural uranium. For the other reactors uranium must be enriched to 2–3% of U-235. So far the United States and the Soviet Union have had a monopoly on the enrichment of uranium. The three American gas diffusion plants at Oak Ridge (Tennessee), Paducah (Kentucky), and Portsmouth (Ohio) were built by the Atomic Energy Commission (AEC), which is the federal agency for research on the peaceful utilization of atomic power established by President Truman in 1947. Truman appointed David Lilienthal, former head of the TVA, to lead the new agency.

The AEC was discontinued in 1975 and its tasks taken over by the Energy Research and Development Administration (ERDA), which indicates that the United States wants to coordinate research in the field of energy. As before research projects are handed out on contracts to university and industry laboratories. Only a small volume is handled by ERDA itself. The regulating functions of the AEC in the field of nuclear energy were taken over by the new Nuclear Regulatory Commission (NRC).

In Europe, a gas diffusion plant is under construction at Tricastin near Pierrelatte on the lower Rhone River. The participants in Eurodif are, in addition to France, Italy, Spain, Belgium and Iran. The plant is expected to be ready in 1979. A separator plant, which utilizes the difference in mass numbers and thus in atomic weight between U-235 and U-238, will be ready a few years later at Almelo in the Netherlands. It is being built for the Urenco group (West Germany, the Netherlands, and United Kingdom). Nuclear power stations use low-enriched uranium (2–3% U-235) while atom bombs require a higher ratio of U-235 (90%). France for its military program has a special enrichment plant at Pierrelatte.

Two private consortia in the United States, organized around Westinghouse and General Electric, gave up their plans for enrichment plants in 1974. Investments amounting to several billion dollars and competition with the federal AEC, which may already have written off its investments, influenced the decisions. Around 1980 the world may face a shortage of enriched uranium. Several countries in western Europe have turned to the Soviet Union for parts of their need of enriched uranium.

Fuel recovery plants — The used fuel cells cannot be thrown away. They contain plutonium, a fissionable element that does not occur in nature but is formed in the reactor, and uranium. Both uranium and plutonium are recoverable resources that may be used in the breeder reactors of the near future. In addition, the cells contain radioactive cleavage products with a negative value which will remain very dangerous for a long time and therefore must be guarded. After being stored under water in the power plant the used fuel cells are sent to a fuel recovery plant. The transport takes place in large steel-clad lead cylinders.

When the useful products have been separated from the radioactive waste, this is being stored under water for a long time. It will later be transferred into solid, insoluble form, glass or ceramics, for future storage that must not allow spread into nature. Because of the youth of the nuclear power industry, decisions about the final storage of radioactive waste have not yet been taken.

Pre-Cambrian granites and gneisses with few cracks and salt mines have been mentioned as ideal storage places. In the United States plans exist for storage in underground salt deposits in New Mexico. Earlier plans for using the salt mines of more densely populated Kansas had to be abolished after negative public reactions. In northern Europe suitable rocks for storing radioactive waste are common in Sweden and Finland.

The Future of Nuclear Power — Professor Leslie and others have drawn the conclusion that the switch to nuclear energy should take place as soon as possible since economically it is competitive with energy from fossil fuels. Fossil fuels have alternative uses, and should be saved. Uranium can only be used for energy production. Coal and oil are 'dirty' fuels. They are used with obvious risks to the environment. Uranium is clean. Since Svante Arrhenius in the beginning of this century drew attention to the possible longterm effects of coal and oil burning and the consequent increase of the atmospheric carbon dioxide content on the climate of the earth this issue has from time to time been hotly debated. More recently attention has been drawn to the emission of sulphur dioxide and the increased acidity of soils and lakes. No such risks exist with nuclear energy. The coal and oil cycles also entail risks to workers and environment at the mining and transportation level that has no counterpart in the uranium cycle.

The fear of nuclear energy is primarily a fear of the atom bomb. To fully understand the operation of a light water reactor and of an atom bomb requires knowledge of nuclear physics far beyond the scope of this book. We must therefore refer the reader to textbooks in nuclear physics on this point. However, for the further argument one statement is important: a lightwater reactor cannot blow up like an atom bomb. The American Rasmussen report shows that a lightwater reactor is the safest place of work in industry.

The connection between peaceful nuclear energy and nuclear weapons is another question. Fissionable material from a fuel recovery plant can be used in the construction of nuclear weapons. But a state or similar organization that wants to make nuclear weapons can build those even without a program for nuclear power production. Mankind cannot unlearn the knowledge of nuclear physics. The problem of avoiding nuclear warfare is a separate issue, which would have to be faced even if mankind were to abstain from using nuclear energy. Under those circumstances it is not realistic to envision that the nations of the world would agree not to use nuclear reactors to produce part of their electricity, even if it could be shown that the reserves of fossil fuels were sufficient to bridge mankind over to the period of inexhaustible energy sources (fusion energy, concentrated solar energy). The oil crisis demonstrated that global energy reserves do not guarantee national or regional (continental) security in the field of energy. With more energy options, nations can obtain the desired security.

The period of rapid expansion in the field of nuclear energy (fig. 4) started around 1970. It will entail hardships, primarily the difficulty of accumulating enough capital (cf the period of water power expansion in some countries). With a global addition of 95 GW nuclear capacity a year in 1985 capital costs for the power stations will be US $38 billion to which comes the cost of fuel production. A speeding up of the already high investment tempo in the nuclear power industry as a result of the oil crisis (cf France) will strain the capital market.

The switch from cheap petroleum to equally cheap but more capital intensive nuclear power will be to the disadvantage of the poor countries where capital shortage is even more pronounced than in the industrial countries. In the poor countries the interest rate for longterm capital, so important in the cost estimate of the nuclear power station, is much higher than in the industrial countries. Experience from the exploitation of the water power potential in countries with different GNP/capita confirms this assumption, The poor countries have hardly started to utilize their water power potential.

Difficulties through delays in the construction of nuclear power capacity may also be caused by confrontations between environmental protection groups and the nuclear power industry. This confrontation, however, seems to be less pronounced in countries where nuclear power plays a large role, like Britain and France, than in countries where it is of a more subordinate role, like the United States and Sweden.

Great Britain – The pioneer among the British nuclear power plants, Calder Hall in Cumberland, went on stream 1956–59. It was the first full scale nuclear plant in the world. Great Britain has played a prominent part in the field of nuclear energy. In 1971 it had more nuclear energy in its electricity production (11%) than any other country. Until 1970 Britain was the leading producer of nuclear energy in the world. A number of research centers are scattered over the country. Research and development is under the control of the Central Electricity Generating Board (CEGB).

The British nuclear power stations are located around the coast. The economies of scale are obvious and new reactors become larger and larger. Dungeness on the Channel, has two reactors with a combined capacity of close to 2 million kilowatt. Other large stations are Wylfa in Anglesey and the Durham plant in Hartlepool. Among older stations are Dounreay north of Aberdeen, Sizewell and Bradwell north of the Thames Estuary, Hinkley Point; Berkeley and Oldburg on the Bristol Channel and Severn, Trawsfynydd on the northern west coast of Wales, Chapel Cross on the Forth of Solway and Hunterston west of Glasgow. Plans in the early 1970s saw a doubling of capacity by 1975 with 13.4 million kilowatt and a capacity of 75 billion kWh.

The comprehensive British nuclear research did not lead to commercially viable reactor types. Nor did the French attempts. Magnox is found in some power stations (e. g. Wylfa and Berkeley) but it is expensive and not competitive with the American light water reactors. It uses natural uranium baked into a magnesium alloy. The reactor is cooled by carbon dioxide and moderated with graphite. A further development of the magnox is the AGR (Advanced Gas Reactor) which works at higher temperature and pressure and as fuel uses uranium contained in stainless steel. The British have spent much money on this reactor but it is as expensive as the Canadian CANDU and uses enriched uranium.

Tab. 20 – Nuclear Energy in the Total Electricity Supply of Selected Countries

Country	Nuclear energy, 10^9 kWh				Electricity 1973, 10^9 kWh	Nuclear energy, %			
	1963	1967	1971	1973		1973	1980	1985	1990
Canada	0.1	0.1	4.0	14.3	262.3	5.5			
United States	3.2	7.7	37.9	83.3	1947.1	4.3	24		41
France	0.4	2.6	8.7	14.0	174.1	8.0	25	45	
Germany (W)	0.1	1.2	5.8	11.8	299.0	3.9		45	
United Kingdom	6.5	24.2	26.9	28.0	282.1	9.9			
Italy	0.3	3.2	3.4	3.1	145.5	2.1			
Sweden	–	0.1	0.1	2.1	78.1	2.7			
India	–	–	1.2	2.2	75.5	2.9			
Japan	0.0	0.6	8.0	9.5*	470.1	2.0			
World	10.6	42.4	106.7	181.3	6042	3.0			

* Refers to 1972.
Source: SY 1974.

In 1974 the British minister of energy decided that the second phase of the British nuclear power program should be based on the domestic SGHWR-reactor (Steam Generating Heavy Water Reactor). It is a hybrid between the AGR and the PWR reactor or between a heavy water and a light water reactor. SGHWR is somewhat more expensive than the PWR/BWR reactors but is safer which should be expected from a heavy water reactor. CEGB voted in favor of the PWR reactor.

Germany (W) − The situation for nuclear power research was unfavorable in Germany after the Second World War but at the middle of the 1970s Germany (W) had a comprehensive nuclear power program. Ten reactors at nine places in the beginning of 1975 accounted for a capacity of 3.5 million kilowatt or 5% of the national demand for electricity. Under construction were 13 reactors with a capacity of 13 million kilowatt. These 23 reactors are in 20 places, 11 of which are on the Rhine or its tributaries. Two plants (Brunsbüttel and Stade) are on the Elbe Estuary and one, Esenshamm, on the Weser Estuary; all others are on rivers in the inland. With such small recipients the waste heat problem will be a major technical issue facing Germany (W). In a densely populated country like Germany it should be possible to find uses for this heat in cities, industries, and aquiculture. The largest power plant, at Biblis between Karlsruhe and Darmstadt, is expected to have a capacity of 5 GW in 1985. Germany (W) is expected to have a capacity of 20 GW nuclear power in 1980 and 45 GW in 1985.

France − Because more than Germany (W) and Britain it had allowed its coal industry to lag, France was more severely hit by the oil crisis. It has a very ambitious nuclear power program. During the two years 1976−77 France will build 12 reactors, each with a capacity of one million kilowatt. After that reactors will be even larger, 1.3 GW.

The French reactor market has been dominated by two companies producing the American BWR and PWR reactors on license but in 1975 production was concentrated into one company, Framatome, in which the American Westinghouse has a minority interest. France in the future will concentrate production to pressurized water reactors (PWR) which the French have found to be safer than the boiling water reactors (BWR). The government plans to have a purely French species of the PWR before 1982 when the license agreement with Westinghouse expires.

The other reactor company, Cie Générale d'Electricité, associated with the American General Electric, will concentrate on the turbine market through its subsidiary Sté Alsthom. According to plans the two companies will cooperate in offering turn key nuclear power plants in the world market.

United States – The first American nuclear power plant was the small pile at Shippingport, Pennsylvania, that went on stream 1957. It was soon followed by reactors in Illinois (1960, Dresden near Chicago, the first full-scale BWR, 180 MW), Massachusetts (1961), New York (1962, Indian Point, Buchanan, the first full-scale PWR, 150 MW), and California (1963). The largest of the five reactors had a capacity of 265 000 kW. New ones rapidly became larger and the total nuclear power production increased at a high rate. The definite commercial break-through came around 1968. At the end of the year 1970 nuclear power accounted for 2% of the American electric energy. It is expected to reach 24% by 1980 and to exceed 50% by the year 2000.

The American energy crisis of 1972, which in contrast to the oil crisis of 1973 was national in scope, partly resulted from delays in the nuclear power program. Some reactors were retarded up to three years because of actions from environmental protection groups. As shown in table 20, the United States in 1973 was behind Britain and France in the share of nuclear power in its electric energy output.

Like other industrial nations, America in the mid–1970s was in the middle of a rapid expansion of its nuclear power capacity. However, recession, shortage of long-term capital and energy saving as a result of higher energy prices (the actual demand was much lower than projected), led to delays in the construction start of many reactors. Possibly, the United States and other industrial countries have now entered the period of decreasing increase rates for energy demand that must eventually come (fig. 4). The decrease in increase rates may be larger than already foreseen in the long-term projections, which as a rule were based on lower energy prices.

The increase in reactor size is reflected in the American construction program. In 1974 the US reactors were 50 with a combined capacity of 33 GW. For another 58 reactors with a capacity of 57 GW building permits had been obtained. Applications had been submitted for 80 reactors with a capacity of 89 GW.

Sweden – A small heavy water reactor built to heat some suburbs south of Stockholm was put in operation 1963 at Ågesta on Lake Magelungen near Farsta. The construction of a full-scale heavy water reactor for production of electricity (140 000 kW) at Marviken east of Norrköping was discontinued in 1970, not being competitive with the light water reactors. Some of the buildings at Marviken were used for a conventional thermoelectric plant, popularly known as 'the only nuclear power plant in the world using fuel oil'.

A light water reactor of BWR-type is now produced by ASEA-Atom at Västerås. Sweden's nuclear power production will be concentrated to four places, at least until the 1980s: Simpvarp north of Oskarshamn with two reactors (440 and 580 MW) in operation and a third (900 MW) planned for 1979;

Ringhals north of Varberg with two (760 and 820 MW) in operation and two more (900 MW) planned for the late 1970s; Barsebäck north of Malmö with two reactors (580 MW) going on stream in the mid – 1970s; and Forsmark north of Stockholm with two units (900 MW) being ready by the late 1970s and two more in the 1980s. Heat from Barsebäck will be transported to nearby cities.

USSR – After the inauguration of her first small test pile in 1954 it took a long time before the Soviet Union went on with her nuclear program. In 1958 a nuclear power station was opened somewhere in Sibiria. It was designed to produce plutonium for nuclear weapons and is reported to have had a capacity of 600 MW. The first full-scale civilian station came 1964 at Belojarski 40 km east of Sverdlovsk. It had an initial capacity of 100 MW to which new units were added later. Also in 1964 the Novovoronesjski plant on the Don south of Voronesj was opened with one reactor (210 MW). In the early 1970s it had been expanded to 1500 MW. In the late 1960s the Soviet Union produced a series of reactors of 440 MW capacity. Two of these were for a station south of Murmansk and two more for one near Yerevan. In 1975 the largest Soviet nuclear station was in Leningrad (2000 MW). An even larger one is under construction at Kalinin between Leningrad and Moscow (4000 MW). Reactors here are of larger capacity (1000 MW). The total nuclear power capacity of the Soviet Union is planned to increase 8–10-fold between 1975 and 1985.

Nuclear power is ideal for large energy needs in sparsely populated areas far from conventional sources of energy. One kilogram of uranium in a nuclear power station yields as much heat as 20 tons of coal; in the more efficient breeder reactor it will yield as much as 1500 tons of coal. Transport costs for the fuel which in the early stages of the industrial revolution completely dominated the location of industry and which may still be of importance for the location of some industries, is of no significance for the location of nuclear power stations. On the other hand, the transmission cost for electricity will be an important factor, since nuclear power stations have considerable economies of scale. Not only costs but also the environment must be taken into consideration: transmission lines and transmission swaths from such stations will be much larger than those from present power stations.

At Sjevtjenko in the arid oil field of Mangisjlak east of the Caspian Sea the Soviets have built the first full scale breeder reactor in the world with the double aim of producing electrical power (150 MW) and steam for a desalination plant. However, American satellite pictures have indicated a fire in this station. According to Soviet reports the fire damaged the turbine section and not the nuclear reactor.

India – The ambitious plans for construction of nuclear power capacity have only partly materialized. The Indian government hopes eventually to be able to use thorium as fuel but a prerequisite is a fast breeder reactor. India has the largest thorium deposits of the world in the monazite sand of Kerala.

Existing nuclear power stations use uranium from deposits in Bihar. The large nuclear research center at Trombay east of Bombay employs over 2000 scientists. Further to the north, at Tarapur, is the first nuclear power station of India (380 MW). The first reactor was started in 1969 followed by a twin unit, both of BWR-type. Southwest of New Delhi at Rana Pratap Sagar is the second station with two reactors of the Canadian model (Candu), each with a capacity of 200 MW. A third station with two somewhat larger Candu units (220 MW) at Kalpakkam near Madras has been planned for 1975 and 1976. The Candu reactor has the advantage of using natural uranium, which means independence of enrichment plants, but the disadvantage of higher investment costs than the BWR and PWR reactors. The Candu reactors were thought to have provided India with the plutonium for her nuclear blasts which led to strained diplomatic relations between Canada and India. The United States supplied the heavy water needed in the Candu reactor.

In spite of India's high scientific standard in the field of nuclear power, implementation of plans runs into the consequences of underdevelopment: roads, railways and bridges are not built to carry reactor components which may weigh as much as 200 tons. Western scientists are working on plans to develop zeppelins for such transports. They would simplify the planning of nuclear power plants, not the least in the Third World.

Breeder Reactors

The breeder or fast reactor is called fast because the speed of neutrons on the average is 5000 km/sec, or 1000 times faster than in the thermonuclear reactor, and breeder because more than one neutron from each atom fission causes a new fission. Fast neutrons 'breed' in this sense. Slow neutrons, also called thermal neutrons since they are in thermal equilibrium with the surrounding atoms, can be used for nuclear fission of uranium-235 while uranium-238 and thorium require fast neutrons.

Experimental breeder reactors have been built in the United States and several other countries. Such reactors are needed if fissionable materials (uranium and thorium) are to be more than a short parenthesis in the history of energy. Only breeder reactors can utilize thorium and the common uranium isotope 238 which makes up 99.3% of natural uranium. A breeder reactor will make available some 50−70% of the energy of uranium (some will be used in the breeding process) while a thermal reactor will make use of only 0.7%. A breeder reactor meeting the safety requirements of modern society would thus dramatically widen the raw material base of the electroproducing industry. However, in practice the efficiency will be lower than 50−70 percent since there is a tradeoff between efficiency and safety. Safety on a par with that of light-water reactors or better will yield considerably lower percentages of energy, say, some 40 percent. Since data are lacking, statements about the economy of breeder reactors can not yet be made.

Research on breeder rectors has been concentrated primarily to types cooled with liquid metal, liquid metal (cooled) fast breeder reactor (LMFBR). It uses liquid sodium for cooling and heat exchange and plutonium oxide as a fuel. An alternative type that so far has met with less interest is the gas cooled fast breeder reactor (GCFBR) in which helium is used for the heat exchange with the steam generator. Since the heat capacity of the gas is less than that of the liquid metal the gas must be compressed to pressures of 70 to 100 atmospheres and the whole reactor built with a shell of prestressed concrete. Germany (W) expects to have such a high temperature reactor in operation by 1977 at Schmehausen (300 MW).

It is widely thought that France may be closest to a commercial breakthrough with a LMFBR. The prototype reactor Phénix at the atomic center of Marcoule in the Rhone Valley was started in 1974 and after almost two years of operation had not run into any serious problem. A full-scale breeder reactor (1200 MW), the Super-Phénix, is under construction. Also the British have advanced plans for commercial breeder reactors of 1300 MW.

The scientists who at the end of the 1940s held out the promises of nuclear power had the breeder reactor in mind. The first electricity produced with nuclear energy was obtained from an experimental breeder reactor in Idaho in 1951. This technique in the mid–1970s is much more advanced than the fusion technology 'which still remains to be invented'. But ERDA in its long term plan of 1975 for energy development in the United States no longer holds on to 1987 as an American target date for commercial breeder reactors. The Clinch River reactor at Oak Ridge (Tennessee), to be built 1976–82 and tested for five years, was to provide the data. ERDA in its program also emphasizes research on solar energy as an alternative to breeder reactors and fusion power within the category of almost inexhaustible sources of energy. Electricity produced with solar energy may become of interest to the United States and other low-latitude countries with large deserts but hardly to northern Europe with its long winters.

Fusion Energy

The first steps towards a controlled utilization of hydrogen power for peaceful purposes have been taken along two separate lines. In both cases the Soviet Union has made important innovations. Research has primarily been concentrated to magnetic containment of fusion but also to experiments with laser fusion. Most informed scientists seem to be of the opinion that fusion energy at the earliest will be available at the end of the century.

The fuel for fusion has to be sought among the lightest atoms. Interest has been focused on the two heavy isotopes of hydrogen, deuterium and tritium. Heavy water (deuterium oxide) is found in all water in concentrations of 200 ppm while tritium is obtained through neutron bombardment of lithium in reactors. A jug of water contains enough heavy hydrogen to supply a family with energy for a year; the energy of the seas amounts to $10^{10}Q$. Fusion of a deuterium with a tritium nucleus (the D-T-reaction) is expected to be the first solution to the fusion problem. It occurs at a temperature of some 100 million degrees. The fusion of two deuterium nuclei (the D-D-reaction) requires higher temperatures and will be more difficult to achieve. As lithium is a rare metal, the D-T-reaction will not provide unlimited energy, probably less than the breeder reactor.

In the early 1950s no one knew that it would take a long time to build a fusion reactor. Few scientists were aware that first a new branch of science, plasma physics, had to be developed. At temperatures of several tens of million degrees required for fusion (thermonuclear energy) atoms can no longer exist; matter becomes a plasma of free nuclei and free electrons. Plasma is the fourth form besides solid, liquid and gas. Important contributions to plasma physics were made by, among others, the 1970 Nobel laureate in physics, Hannes Alfvén.

The most promising model for magnetic containment of fusion is the toroidal or doughnut-shaped tokomak, a Soviet design, in which a plasma of fast, free electrons and nuclei move in one direction in a magnetic field.

In one of several American tokomaks, stationed at the University of Texas, a record temperature of 200 million degrees was reached in 1974, but only for less than a hundred millionth of a second. This was seen as an important step towards controlled fusion power but no breakthrough. In Europe this type of research is carried on within EURATOM, the nuclear research organization of the EEC.

Water Power

Water power is kinetic energy or the energy of a body in motion. It equals half the product of its mass and the square of its velocity $\frac{mv^2}{2}$. The opposite is potential energy.

During millenia before the advent of the modern electroindustry, running water was utilized to turn water wheels, the location of which decided the exact sites of the pre-industrial manufacturing plants (mills, tanneries, iron works and so on). Manufacturing industry was then primarily located in the countryside. Even small creeks were utilized for manufacturing purposes.

With the steam engine, industry became concentrated to cities, primarily on or near coal fields and in coal-importing harbors. Large steam engines for multistoried factories were more economic than small ones. Economies of scale led to a concentration of manufacturing industry and of the urban population. The industrial belts of Europe and North America with their grimy factory towns mushroomed within a few decades. Northwestern Europe, primarily Britain, and northeastern United States became the workshops of the world.

Electricity introduced a decentralizing element. The utilization of running water for the production of electric energy was concentrated to a few, large power stations. The turbines could also be turned by steam produced with fossil fuels and more recently by nuclear power. In many industrial countries water power only accounts for a small part of the total output of kilowatthours. In contrast to earlier forms of energy, electricity, after a short initial period, could be transmitted over longer and longer distances at low costs. Electric energy (and the internal combustion engine) has permitted a decentralization of production and settlement. Factories and wholesale warehouses no longer had to be located at ports and railroad stations but could be located far out in the suburbs of metropolitan areas and in medium-sized and small cities.

The running water turning the turbines of the power plant is part of the hydrological cycle which derives its energy from the hydrogen fusion in the sun. The sun's rays are also, by way of photosynthesis, the energy source of wood and fossil fuels. However, economically there is a great difference between water power and fuels originally formed in the plant and animal realms. Evaporation from the water surfaces of the earth (seas, lakes and rivers) and from the biosphere (plants and animals) is continuous and precipitation is frequent. The water that turns the turbines is continuously being replenished.[17]

Other forms of kinetic energy may also be considered: waves of the sea, tides, and sea currents. A feasible technique for utilizing these sources is not yet within sight. The first tide water plant was opened in 1967 in the Rance estuary between Dinard and Saint Malo in Britanny. The Russians have constructed a small experimental plant at Kislaya Guba north of Murmansk. They have plans for a large plant at the entrance of the White Sea. The British plan a tidewater plant in the Severn estuary at Bristol and the Canadians one in the Bay of Fundy between Nova Scotia and New Brunswick. A special form of water power, that theoretically might be considered for the future, is thermal energy or the vertical differences of temperature in the sea.

Hydroelectric power stations — Since energy from running water is a function of the mass and the speed of water, climatology, geomorphology and hydrology are obvious basic sciences for the study of the water power potential. Mountains near the west coasts of continents in the temperate west wind belts have the largest and most evenly distributed precipitation and thus the highest potential on these latitudes. Examples are the northern parts of the west coast of North America, southern Chile, and Iceland and the Scandinavian mountains as well as the Alps. Even larger rainfall is registered for areas with convection rain near the Equator and with monsoon rains in Asia. The equatorial regions of western and central Africa have the largest potential water power resources of the world. Only a very minor part have been utilized. The Western Ghats of India, Himalaya, the border mountains between India and Burma (with the record station Cherrapunji) and Japan are examples of regions potentially rich in water power. But only Japan in Asia has built out much of its water power potential.

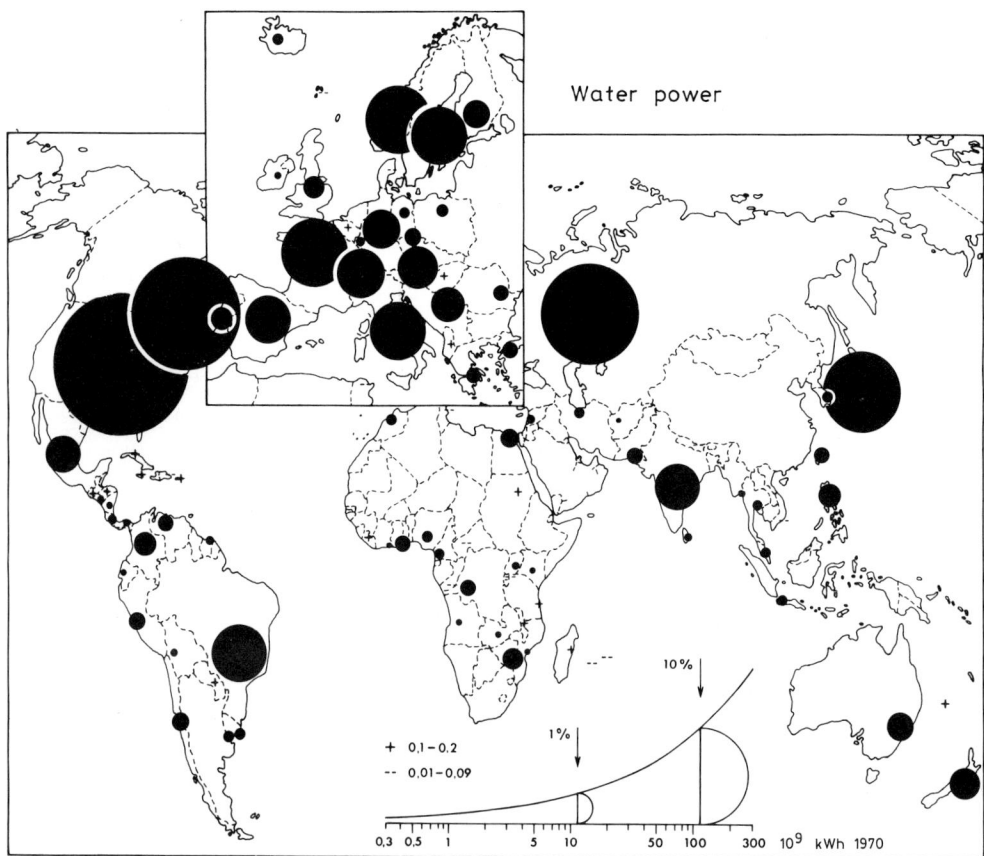

Fig. 20 – Hydroelectric power plants are capital intensive. Access to cheap capital (measured by the interest rate for long-term loans) is more important than the physical geography for the construction of such plants. For water power, the global production map shows little resemblence with a map of global potential. The industrial countries are over-represented and the tropical regions, with their vast potential, grossly under-represented.

In cost comparisons of water power and other forms of electric energy, the interest rate for long-term capital weighs heavily in the water power table, even more than in the table for nuclear power. Regions with a large water power potential as a rule are far away from the concentrated market areas. Transmission costs must therefore be added to the water power table to achieve comparability. Multinational corporations with access to long-term capital at reasonably low interest rates have been active in the exploitation of peripheral water power sites in e. g. Norway and Canada but only to a very limited extent in tropical Africa (e. g. the Volta-project in Ghana).

Until the mid-1970s water power accounted for 100% of the electricity consumption in Norway. But Norway is an exception among the industrial countries. As a rule, water power in those countries accounts for a couple of tens of percents or less, table 22. United Kingdom and Germany (W) have very low shares, only a few percent, while Sweden has $^2/_3$. The Alpine countries France and Italy also have large shares, about 40%. In tropical countries, which utilize only a very minor part of their water power potential, hydroelectric power often accounts for a high percentage of their small supply of electricity.

An alternative to transmitting electric power to distant markets is the location of energy intensive industries at large hydroelectric power stations. The electrolytic refining of some metals like aluminum, magnesium, titanium, and zinc and the making of ferroalloys, heavy water and ammonia are examples. For Norway and Canada such plants are important export industries, a way of exporting cheap water power.

Electric Energy

From year to year electric energy accounts for an increasing share of the energy consumption of the world. Already extrapolators of curves are talking about an all-electric society in the first half of next century. While the total energy production of the world in the 1960s had an annual growth of 5%, the production of electricity increased by 7.9%, see table 21. Hydroelectric power plants and nuclear reactors only produce electricity, which thus is their primary energy. Thermoelectric plants use primary energy (coal, oil, gas) as an input and their output, electricity, is secondary energy. This complicates all comparison, see table 22.

In the 1950s and 60s energy prices were low and decreasing in the industrial countries but this trend was changed a couple of years before the oil crisis in 1973. In the United States the average real price of electricity increased in 1971 which, with one exception, was the first time since 1946. As seen from figure 21

Tab. 21 — Annual Rate of Increase in the Production of Electricity, per cent

Decade	US	UK	Germany (W)	France	USSR	Japan	Sweden	Norway	World	Total energy, world
1950s	8.5	9.3	10.2	8.1	11.3	9.7	6.7	5.7	9.3	5.1
1960s	6.9	6.1	7.6	6.9	9.7	11.8	5.7	6.4	7.9	5.0

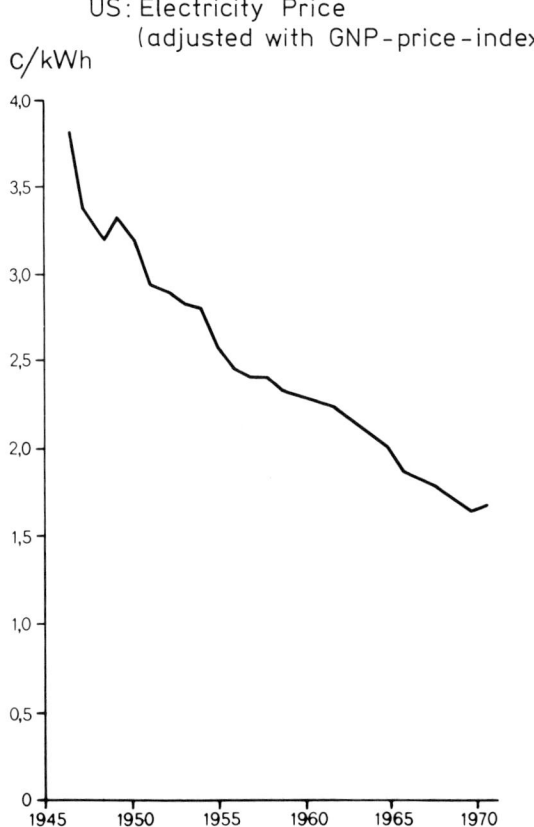

Fig. 21 — Prices for electricity in the United States 1946–1971, corrected for inflation. The price went down all years except 1949 and 1971.

the average kilowatthour-price in the United States declined from 3.8 c to 1.6 c in that quarter-century. Electricity consumption in the world increases faster than other final energy consumption with one exception, natural gas. The electricity consumption of the world increased tenfold between 1940 and 1970.

On the average consumption in the industrial countries doubled in less than ten years (which corresponds to an accumulated increase with 7.2% a year). Growth in other countries was even faster.

The American Federal Power Commission (FPC) in the early 1970s expected the consumption of electricity in the United States to double in the 1970s to $3 \cdot 10^{12}$ kWh and after that again to double in the 1980s. Those estimates were extrapolations of past trends. But projections made before the oil crisis may turn out to be unrealistic. Econometric sensitivity analyses show demand to be influenced by the price as it should be according to economic theory. A doubling of the real price before the turn of the century would mean that the American demand for electricity would increase by only 33% over 1970, i. e. an annual increase of only 1%. The longterm development of the price for energy will strongly influence production.

In the early 1970s most experts thought that energy prices would go up. (The unilateral decision by OPEC in 1973 to charge a very large monopoly rent on oil which in a few months increased the world market price four-fold is not considered here. Such actions may in the short term influence the world balance for energy and finance but should not influence the long-term price for energy.) A warning is therefore appropriate against using exponential growth when predicting the most likely future production curve for electricity. Table 21 shows for the 1950s and 60s declining rates of increase in most countries for that period of decreasing kilowatthour prices. With increasing or stable prices the decline should be more marked. The rapid population increase, strongly concentrated to the tropical lands, and rising expectations about higher standards of living in most countries, especially in the poor nations, should exert a strong upward pressure on the production curve. However, demographic data indicate that the world with the 1970s enters a period of a rapid decline in the population increase also in the countries of the third world. The industrial countries are already very close to mere reproduction levels (net reproduction rate 1.0 or less). With lower population increases it will be easier to satisfy demands for higher individual energy standards even within moderate production increases.

A serious objection against electricity produced with fossil fuels in condensation works is the low efficiency, just over 30% against more than 70% when gas and oil are used directly in burners in homes and industry. The efficiency in the American electric power stations increased from 5% in 1900 to 33% in the early 1970s. With smaller power stations built according to the counter-pressure principle adjacent to cities it is possible to use the waste-heat of the power station in the central-heating systems of the city and thus to increase the efficiency of the power station to 55–80%.

Efforts to increase the electricity-producing efficiency of the condensation works are up against the laws of thermodynamics. It can be done only by

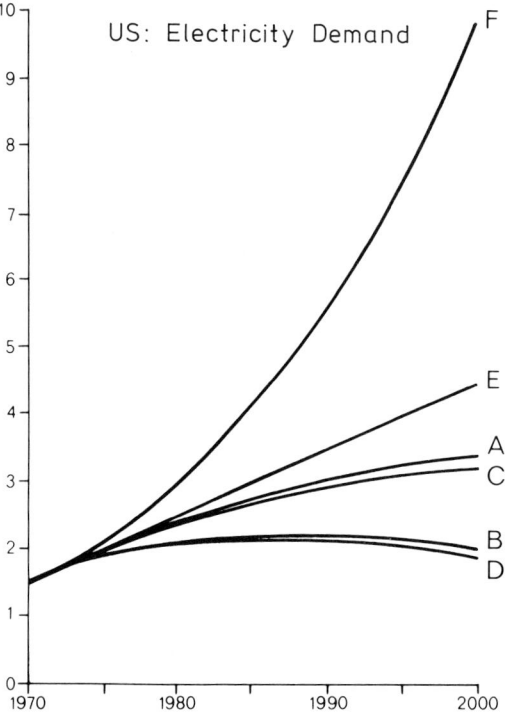

Fig. 22 — Projections of electricity demand in the United States made around 1970 under various assumptions. The Federal Power Commission (FPC) assumes a doubling in the 1970s and again in the 1980s. The projection (F) is a simple extrapolation of the recent trend. The period was one of relatively rapid population increase and declining prices for electricity (fig. 21).

D. Chapman, T. Mount, and T. Tyrrell in a critique of FPC introduce variables that may influence the demand for electricity: price, population increase, income and price of natural gas. An econometric sensitivity analysis shows the price of electricity to be the deciding variable, much more important than the rate of population increase. The price of the competing natural gas does not seem to play any significant role.

All curves except C and D are based on the same assumption about a continued population increase in the United States while C and D assume population increase zero from 2035. This, in turn, assumes that a net reproduction rate of 1.00 is reached in the 1970s. The observed demographic datum for 1975 was below that assumed in C and D.

For a doubling of the demand every ten years (F), prices should be reduced by 50% before the year 2000. With stable prices, demand increases more slowly (E). If prices are doubled by the year 2000, electricity consumption will climb with only 33% in 30 years (B).

Assuming a small price increase, the FPC arrives at an alternative curve (A). Combined with the lower population increase, the alternative curve will be only slightly lower (C). The same holds if the B is combined with the lower population increase assumption (D).

Source: "Electricity Demand Growth and the Energy Crisis", Science, nov. 1972.

increasing the temperature of the incoming fuel or lowering the temperature of the cooler. Efficiency is a function of the difference between the two. A higher efficiency can be achieved in a combined gas and steam turbine plant, now being developed in the United States. Hot gas first turns a gas turbine generator and then produces steam for a conventional steam turbine. The combiplant has an efficiency of 40%. Conventional gas turbine generators, developed with experience from the aerospace industry to be used for top loads, only have an efficiency of 25%. If the input temperature can be raised the combiplant will have even higher efficiency, but for such high temperatures difficult material problems have to be solved first. An efficiency of 50% is expected for the early

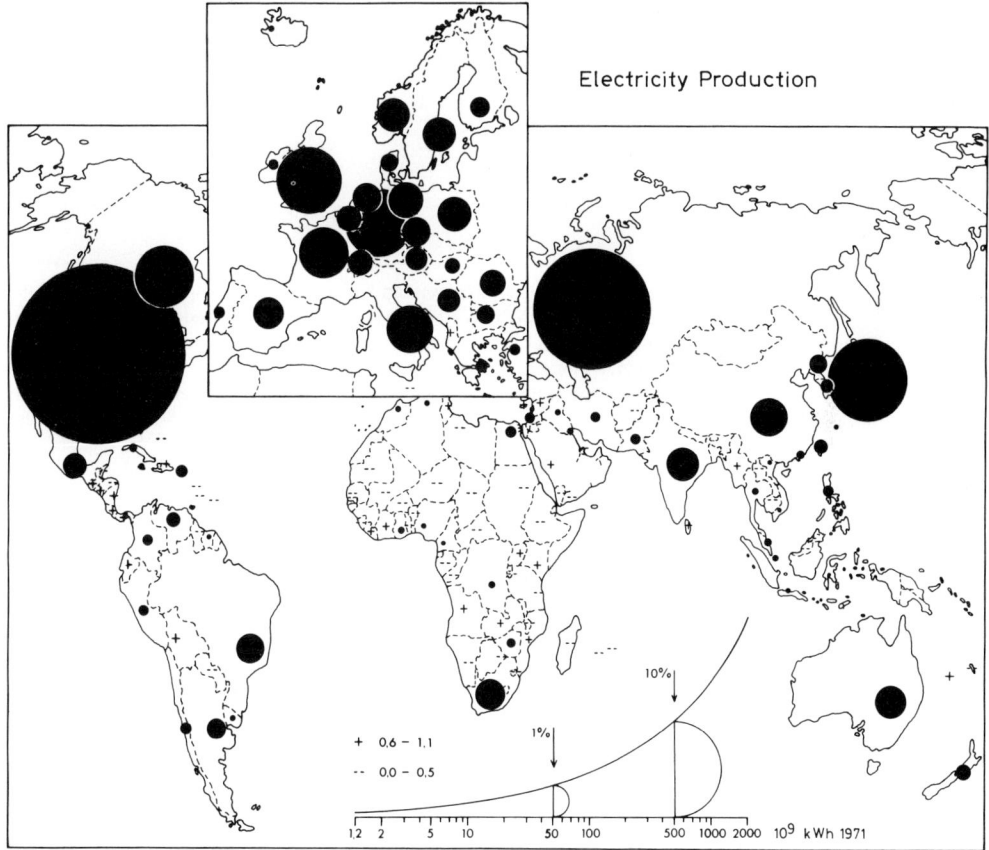

Fig. 23 — The production of electricity is a good indicator of the industrial production capacity of a country. The per-capita consumption of electricity reflects the standard of living. Some industrial countries, with access to cheap hydroelectric power and exports of energy intensive products such as ferro alloys, aluminum, nitrogenous fertilizers, or heavy water — for instance Norway, Switzerland, and Canada — will be somewhat overrepresented.

Tab. 22 – Electricity Consumption and its Share of Total Energy Consumption in Selected Countries 1970 (10^9 kWh)

Country	Electricity consumption (1)	Total energy (2)	Nuclear power (3)	Geo-thermal power (4)	Water power (5)	Column 5, as % of 1 (6)	Column 1, as % of 2 (7)*
US	1638	17343	21.8	0.5	250.6	15.2	12.4
UK	248.6	2275	25.4	–	5.7	2.3	15.3**
Germany (W)	237.2	2392	6.0	–	17.8	7.5	13.5
France	140.7	1465	5.1	–	56.6	40.2	11.5
Italy	117.4	1094	3.2	2.7	41.3	35.2	13.3
USSR	740.4	8202	2.5***	–	124.4	16.8	20.3
Japan	359.5	2522	4.6	0.2	80.1	22.3	21.2
Sweden	60.6	386.1	0.1	–	41.5	68.5	18.4
Norway	57.6	142.1	–	–	57.3	99.5	40.8
India	60.0	780.5	2.3	–	24.6	41.0	8.8
Brazil	45.5	341.6	–	–	39.9	87.7	12.3
Zaire	3.19	11.9	–	–	3.15	98.7	26.8
World	4901	52009	79.3	4.1			

* SY (UN) lists total energy in coal equivalents and electricity consumption in kWh. Conversion factor: 1 ton coal = 7600 kWh. Water power, nuclear power and geothermal energy are primary sources of energy which can be added to the fossil fuels. Electricity from thermal power stations is secondary energy; coal, oil or gas were used as fuel. Primary energy used for this production was arrived at by multiplication with 3 (assumed efficiency 33, which is on the high side). The estimated input of fuel was then deducted from the total energy before the percentage of electricity was computed.

** UK Atomic Energy Authority has estimated the share of electricity in the total energy consumption and for 1970 arrives at a higher value, 16.3%. The share grew from 0.8% in 1920, 2.1% in 1930, 4.2% in 1938, 7.5% in 1950 to 13.0% in 1960.

*** Refers to 1968.

Source: SY 1972.

1980s at a temperature of some 1440 °C for the incoming gas. A major advantage of the gasification technique in electric power production is the removal of sulfur from sulfuric fuels before the fuel enters the power plant.

Another technique of producing electricity with an input of sulfuric fuels without causing air pollution is the magnetohydrodynamic (MHD) generator, which is on the experimental stage at many places, but not yet in operation. This generator has no moving parts. The conducting gas or fluid moves through a pipe lined with electrodes and surrounded by coils that produce a magnetic field across the pipe. This generator produces direct current and is very efficient, especially if combined with a conventional steam turbine. The temperature of the fuel is very high, over 2000 °C, which is necessary for the ionization of the gas.

In all countries electricity accounts for an ever increasing share of the final energy consumption. In Great Britain the cost relation for industrial users of oil, coal and electricity expressed in delivered energy (kWh) was 1 : 1 : 7 in the early 1970s. Fifteen years earlier it had been 1.3 : 1 : 10. With an increasing share of nuclear power in the national electricity mix the price for electricity should be less sensitive to increases in fuel prices. A doubling of the import price for oil would lead to an 80% increase in the electricity cost. For the same cost increase in electricity from a nuclear power plant the uranium price would have to go up twentyfold (thermic reactor) or a hundredfold (breeder reactor).

As shown in table 21 the consumption of electricity has doubled every eight to ten years. This growth rate can be extrapolated (and verified) backwards to the First World War and is often extrapolated forwards. But the table shows a declining growth rate in the 1960s compared with the 1950s for the world and for all but a few countries (among them Japan and Norway). This means that the world presumably had entered a period of declining growth rates in the field of electric energy elready before the oil crisis. The same probably holds true for the demographic development in the world, but this decline in growth rates sets in later, presumably in the mid-1970s for the world but in the 1960s or earlier for the industrial countries and for many underdeveloped countries in Asia and Latin America. The exponential curves of the Forrester Team at MIT and of the Club of Rome could therefore be criticized as unrealistic already with reference to the development in the twenty years before 1970.

Waste heat — When, in the future, production of electricity has doubled some more times the waste heat problem may become serious. An efficiency of 33% in a power plant means that 1 kWh of electricity leads to 2 kWh of waste heat which eventually ends up in the atmosphere (thermal pollution) after moving various ways. A light water reactor dumps some 50% more waste heat into the cooling effluent than a thermal plant using fossil fuels, partly because of lower efficiency, partly because the thermal plant sends some $1/5$ of its waste heat up the chimney. Waste heat discharged to lakes or rivers influences the ecosystem in a way that is negative or positive to man. A rapid, although not exponential, growth in the amount of thermal pollution may have serious repercussions on the hydrological system if countermeasures based on research are not taken.

Recovery of waste heat from nuclear power plants will be feasible if they are located relatively close to metropolitan cities or in densely populated areas (e. g. Germany or Japan). Heating or cooling of houses, increasing the microbic activity in cleaning plants through higher température of the sewage, the heating of greenhouses and fish ponds are among possible uses that may make recovery of waste heat feasible. In high-latitude countries the influences on the ecosystem of the recipients may be positive rather than negative. Among other things, the icefree days on inland waterways may be substantially increased and the

104   Energy

vegetation period prolonged (if the water is first used for irrigation or subsoil heating). But it should be noticed that also small changes in temperature may have large ecological consequences.

Radioactivity – Radiation may turn out to be a less serious problem than waste heat. It is specific for the nuclear power production. Attention has primarily been directed to the electricity-producing unit, but radiation risks are associated with all steps in the fission cycle.

The risks in the uranium mines have been radically reduced with the lowering of the threshold values to less than $1/20$ of those earlier observed. It will take another decade or two before full knowledge is obtained about the adequacy of the present levels. The death risks in the uranium mines as a result of radiation must be seen as a statistically significant increase in the frequency of cancer. In other words, some more deaths among elderly or former miners will be recorded than for mines without radiation. This should be compared with the number of miners of all ages that are killed while obtaining the corresponding amount of energy in coal mines. Coal mining is the most dangerous industry in the United States.

The manufacturing of fuel for light water reactors takes place in two geographically separated steps at which radiation risks are almost absent: the enrichment of uranium and the fabrication of enriched uranium into fuel elements of various shapes.

The operation of light water reactors involves many built in security measures. The extensive American Rasmussen report, published in 1975, established that reactors of BWR or PWR type are the safest places of work in industry. Much discussion has been focused on 'the worst imaginable accident' that could befall such a unit. It can not, intentionally or unintentionally, become an atom bomb. The extraordinary safety of light water reactors now seems to be widely accepted. The discussion instead centers on the next step, the fate of the used fuel elements. As shown earlier they are sent to recovery plants for recovery of fissionable plutonium and uranium and for the final disposal of radioactive waste products.

Other environmental deterioration – The increased energy production also leads to other forms of environmental deterioration, associated with the mining of fuel, and with the power plants and the transmission lines.

The strongest impact on the ecosystem is usually caused by coal mining, above all open pit mining of low-grade coal such as brown coal, not to mention peat. However, the German Rheinische Braunkohlenwerke of Köln has proven at its Fortuna-Garsdorf mine at Bergheim, the world's largest mine with a production of 42 million tons a year, that rehabilitation can be made 'at a profit' and that the refilled land can be made better for agriculture than the original land. At

Bergheim the topsoil was scraped off before mining started, stored and, after the coal had been taken out and the mine refilled, piped with water and fertilizer onto the new land, which was given a favorable 1°.5 slope facing the south. Restrictive legislation proposals in the United States aim at similar favorable results.

Oil from shale and tar sand will also lead to very large mining operations and especially in arid regions may leave bad scars on the landscape. Production of oil on the shelves off densely populated coasts runs into serious opposition, more because of accident risks than because of observed pollution. The opposition against the planned oil exploitation off the coasts of New England and New Jersey is compact but complaints about pollution from existing oil fields off Texas and Louisiana seem to be missing. The waters around the oil platforms here have become popular for sport fishing.

Breeder reactors with uranium or thorium as fuel would lead to minimal inroads on the environment compared to alternative ways of producing electricity. The fuel quantity to produce a given amount of energy would be negligable in contrast to the amounts used at a fossil fuel station.

Exponential growth in the consumption of electricity at the rate of the recent past, with doubling in less than ten years, would soon lead to excessive visual blight where the horizon everywhere would be disturbed by high voltage lines, power lanes or power stations. A leading theme of this book is the irrationality of exponential growth. The consumption of electricity will stop growing long before the nightmare has become a reality. As shown in figure 22, politicians have an efficient control instrument in the price of energy. The rapid consumption increase of the recent past went parallel with a rapid decline in real prices. Already a freezing of the 1970 prices would lead to a drastic reduction in the increase rate of electricity consumption. A premium on energy saving (e. g. subsidies for insulation of houses) would further hold back the increase in electricity consumption in regions with electrically heated (or cooled) buildings.

The long-term consumption platform for energy in the industrial countries may find a lower level than indicated by figure 7 for the United States. It should not be ruled out either that future power stations and transmission lines may be located underground.

Metals and Other Minerals

Geology

Economic geology is one of the oldest forms of human knowledge. The first human beings some 2.0 million years ago were defined by L. B. S. Leakey as users of cutting tools. They soon learnt which stones were better than others to shape for this purpose. Also clay was used early, first for pots and later for bricks. Neolithic man got acquainted with gold and copper. The first used metals were probably found as native metal on the bottom of creeks. Among Egyptians, Babylonians, Assyrians and Indians precious stones were held in high esteem. The Pharaohs sent prospectors and miners to Sinai and the Sudan looking for turquoise and emerald. Lapis lazuli was probably obtained from Afghanistan which points to early distant trade in minerals.

Copper was in general use in Europe around 4000 B. C. and gold as well as copper had been known in Egypt several millenia earlier. Greek and Roman philosophers were interested in ores and their genesis. They took notice of the mines of their time, e. g. the famous silver-lead-mine of Laurion, Attica, on which Xenophon wrote a report in 365 B. C.

Georg Agricola (Bauer), born in the Saxon Erzgebirge, Germany, 1494 and dead in 1555, is credited with the first theory about the genesis of ores. He is referred to as the father of mineralogy and metallurgy. Agricola's work, De re metallica, was a great scientific advancement and influenced later writers. It was translated by Herbert Hoover, later president of the United States. Descartes in his Principia philosophae (1644) saw the earth as a cold star with a hot interior. The ore minerals were pressed upwards and deposited as lodes in the cracks of the outer crust.

In the 18th century German and Swedish scientists, primarily professor Torbern Bergman of Uppsala (chemistry), had lively controversies about hypotheses on the genesis of ores. For more than a hundred years the Mineral Academy of Freiberg in Erzgebirge, founded in 1765, was a world center of economic geology. The professor of mineralogy and geology at the end of the century, A. G. Werner, discarded the theories about an interior origin of metals. He advocated the idea that mineral veins had been formed by precipitation from water percolating down from the primeval universal ocean. This origin was ascribed not only to sedimantary but also to igneous and metamorphic rocks. Werner became the standard-bearer of the Neptunist school. Other scholars showed how the fossil content of sedimentary rocks could be used to decide the relative age of layers and for parallelling the sequence of layers at various places.

Through the influence of G. Cuvier the geological thinking of the time was dominated by catastrophism. It was generally thought that earlier periods had been exposed to more dramatic natural processes than the present time.

Neptunism and catastrophism were critizised by J. Hutton of Edinburgh who was the leading representative of the Plutonist school. It also adhered to the doctrine of uniformism: to understand the rocks Hutton and his disciples only needed the geological processes that currently could be observed. In his Theory of the Earth (1788) Hutton considered igneous rocks including many ores to have been formed when molten magma from the interior of the earth crystallized. Hutton's school laid the foundation for the later development in ore geology.

The big fight between Plutonists and Neptunists led to increased field observations. The Frenchman Elie de Beaumont in 1847 showed that steam played a great role in the formation of ore veins and von Cotta in 1859 in a paper from Freiberg pointed out that there are many types of ores and that they were formed by widely separate processes.

Not until 1974 could the initial phase of the ore formation be studied directly in the field. Geologists and vulcanologists from the French-American Mid-Ocean Undersea Study (Project Famous) went down to the sea bed at a depth of over 2 700 m off Ponta Delgada in the Azores at the rift valley in the middle of the Atlantic Ocean. The crust of the earth, that is separated from the mantel by the Mohorovičić discontinuity, is some 37 km thick in the continents and some 11 km in a normal sea bed, but in the rift valley molten magma is only some ten meters below the surface. According to the geologists, magma gets in contact with water, that under high pressure and high temperature is pressed through the igneous mass. Metals are separated and either deposited in cracks in the rock or gushed out by geysirs and deposited on the seabed, later to be covered by sediments. It is now believed that many ores on land have been formed in one of these two ways, among them the classic copper ores of northern Cyprus. One of the observed geysirs produced almost pure manganese. From a mineralogical point of view this manganese is different from the well-known manganese nodules of the seabed which have quite a different origin.

Iron ore

Ore types – Iron (Fe, ferrum), the basis of our civilization, accounts for 5.0% of the earth's crust. For economic exploitation only concentrations with much higher iron content are of interest. Location and chemical characteristics influence the feasibility and the lower limit for iron content that make the rock an exploitable ore. Few deposits of less than 30% iron content are being used; most large iron mines have ore with iron contents exceeding 50%.

As the non-iron parts of the ore only increase the amount of slag in the blast furnace, the ore is often beneficiated before being charged into the furnace. In the early 1970s more than 50% of the world's iron ore were beneficiated. Beneficiation means that the ore is crushed to ore powder which, however, cannot be used in the blast furnace until sintered to lumps of some 12–50 mm. According to the grain size of the ore concentrate the sintering is made as band or pan sintering (with coarse-grained concentrate) or as pellet sintering (with fine-grained concentrate). As a rule a more fine-grained material is obtained from low grade ores. Beneficiation is a prerequisite for the economic operation of low grade, peripheral iron ore deposits.

The most important iron-carrying minerals that form iron ores are oxides: magnetite (Fe_3O_4), black; hematite (Fe_2O_3), red; limonite or bog-iron ore ($2Fe_2O_3$, $3H_2O$), brown; and siderite ($FeCO_3$), pale brown. Magnetite and hematite are by far the most important sources of iron for the world's steel industry. Pure magnetite contains 72.4% iron, pure hematite 70% and limonite 60%. Siderite only holds 48% iron. The iron-carrying mineral is hardly ever the only one in the ore and the metal content is lower than the theoretical values. Rich iron ores seldom exceed 65–66% iron.

A large part of the world's iron ores are sedimentary. A wellknown example is the limonite that was deposited in lakes and bogs. Under the name of lake or bog ore it played an important part in the early iron industry. Greater current interest have such European ores as the low-grade oolithic siderite ore of Lorraine and its counterparts in England and Germany. The large fields near Lake Superior in the United States also belong to this category. They are made up of hematite ore, made relatively rich (51%) through weathering and natural beneficiation. This ore until recently was the main basis of the large American iron ore production. The mother sediments (taconite), are low grade, strongly cemented and unweathered. They are found in large quantities and after World War II have been mined and beneficiated.

Many sedimentary ores have been changed and metamorphic ores of sedimentary origin may be difficult to distinguish from those of igneous origin.

Iron and steel making – Iron ore frequently carries impurities, some of which may make the ore economically useless, e. g. small amounts of chromium or titanium. Other elements cause an increased amount of slag, e.g. calcium, aluminium and silicon. Sulphur, a well-known problem in steelmaking, is introduced in the iron mainly as a result of sulphurous coke in the blast furnace. This element is removed before the actual steelmaking. Phosphorous is reduced into the pig iron and can not be removed in the blast furnace. The invention of the

110   Metals and Other Minerals

Thomas process or the basic Bessemer process in 1878 was an important innovation in the steel industry of Western Europe. In a Thomas converter phosphorus is removed from the steel through oxidation to phosphate which is tied to the slag. Thomas phosphate is an important fertilizer. The Thomas process opened vast iron ore deposits in Lorraine and Lapland to the European steel industry. A basic open-hearth process (Siemens-Martin) was introduced a few years later.

In the 1960s and early 1970s an intense development of the melt metallurgical processes has occurred in the direction of a more energy concentrated technology. The oxygen processes LD and Kaldo were pioneers in this field.[18] In both processes oxygen is blown from above down onto a steel bath. In the kaldo-

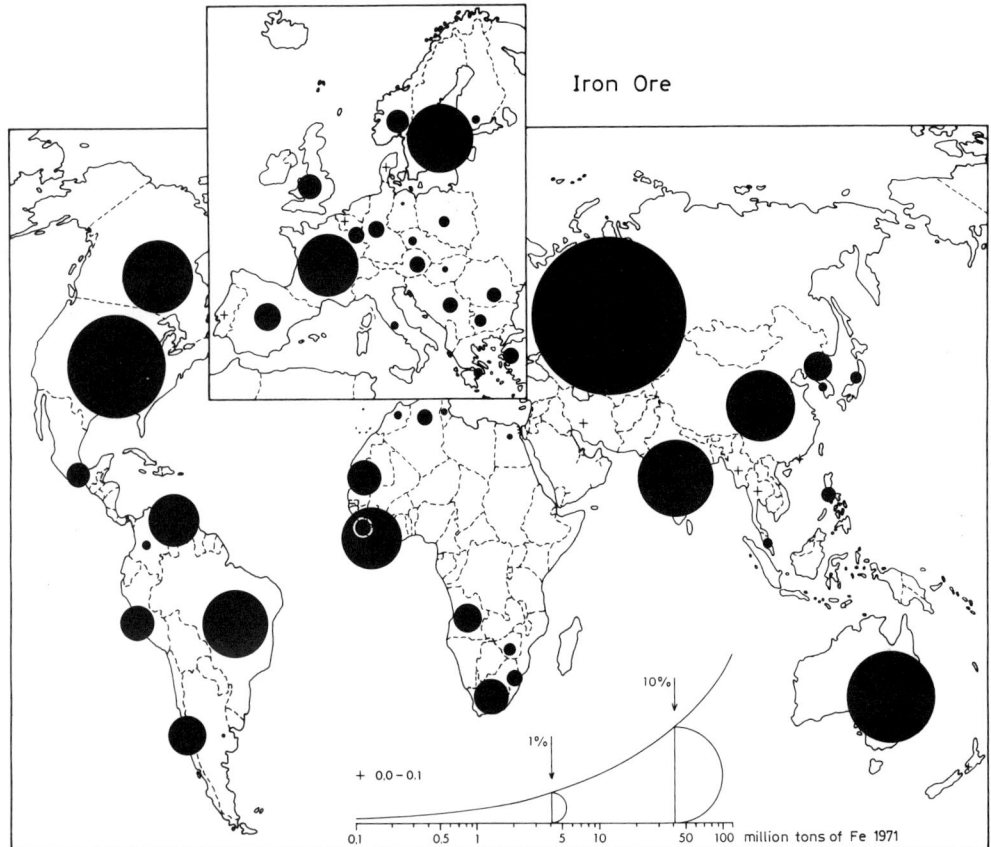

Fig. 24 – In the quarter-century before 1975, transport costs for bulk cargo were radically reduced, not the least for seaborne cargo. Imports of iron ore became possible even from very remote overseas areas. Europe imported ore from Australia and Japan from Brasil. The Swedish ore fields, long enjoying locational advantages with respect to the European iron industry, faced a successively reduced land rent.

process the converter is being rotated. Only LD seems to survive in steelmaking; mechanical problems with a kaldo converter are very real. The largest kaldo converter built takes a 250 tons of liquid steel. The weight of the converter itself is 500 tons and it rotates with 30 rev/min.

The most important of several new processes introduced in the early 1970s was OBM (Oxygen Boden Maximilianhütte). In this process oxygen is blown from the bottom through the bath; through separate inlets around the oxygen nozzle small amounts of natural gas is added to cool the bottom that would otherwise melt. This process and LD are expected to be the most important for blast furnace based steel making in the next ten years. For scrap melting electric arc furnaces are more and more substituting for the slow siemens-martin process.

In the 1950s available iron ore reserves in the world were estimated at more than 100 billion tons containing some 45–50 billion tons of iron. Brazil and India were thought to have the largest reserves, 15 and 10 billion tons of iron ore. But the tremendous deposits of rich ore found in the Pilbara district of northern Western Australia in the 1960s are in themselves said to hold well over 100 billion tons of ore. Other vast deposits have been found elsewhere. Estimates of mineral reserves must always be seen as minimum figures; more careful measuring may prove deposits to be larger than originally thought. New deposits are continuously being found.

North America

United States – The world's leading steel producer since the end of the 1880s, had the largest production of iron ore (Fe content) until 1957 when it was surpassed by the Soviet Union. American output stagnated at about 50 million tons a year while the Soviet production in 1970 exceeded 100 million tons out of a world total of more than 400 million tons.

The American iron and steel industry used to be based primarily on coking coal from the Appalachian coal fields and iron ore from the Mesabi range and other deposits near Lake Superior. In the early stage of the Industrial Revolution (before 1900) steel mills were built along rivers near the coal. Later they were located primarily near the market along the southern shores of the Great Lakes.

The iron ranges in Minnesota, Wisconsin, and Michigan are among the largest hematite deposits in the world, having produced some two billion tons of iron ore. The deposits are pre-Cambrian, weathered sections of sedimentary iron ore beds from which silica has been leached which has caused a concentration of the iron oxides. The Mesabi range has yielded about 80% of the Lake Superior ore. The concentrated ore is found mostly at depths less than 60 m, with a maximum of 270 m. The ore, containing over 50% iron, is scooped up in huge opencuts and hauled by rail a short distance to Lake Superior ports, where it is dumped by gravity into specially designed ore carriers. Until the rapid increase in tonnage and size of oceangoing bulk-carriers after 1955 this was the most efficient transport system for the steel industry found anywhere. Existing locks and entrances to ports before 1970 limited the size of ships to 27 feet draught or 25 000 dwt tons which rapidly made the system obsolete by comparison with the system serving the Japanese steel industry which at the end of the period used carriers of more than 100 000 dwt tons.

The naturally 'beneficiated' ore in the Lake Superior ranges is being exhausted. This became evident during the Second World War and the immediate postwar years when over 50 million tons of iron ore a year was shipped from Duluth-Superior alone and another 20 million tons from Two Harbors. The American steel industry reacted in two ways:

1) By developing methods for utilizing the exceedingly hard rock of the original beds (taconite) which only contains some 25% iron but occurs in enormous quantities. Completely new systems including beneficiation plant, rail cars, loading facilities and vessels were built for this type of ore, which is shipped as pellets with an iron content averaging 63%. For the new taconite trade the Poe Lock in the Soo Canal at the entrance of Lake Superior was opened in 1970. It allows the passage of vessels of 56 000 dwt tons.

2) By prospecting for new ore deposits abroad, primarily in Canada (Labrador, north of Lake Superior), Venezuela and West Africa. Increasing quantities of iron ore (about 45 million tons a year in the early 1970s) have been imported as a result of heavy investments by American ore companies in three continents. American corporations have also been active in Australian projects, whose ore moves to Japan, primarily, but also to Europe.

Alabama has long been a poor second among the American iron ore regions. The ore field is just north of Birmingham where the vast Clinton sedimentary iron ore beds are worked in Red Mountain. The ores, mainly hematite with some limonite, average 37% iron. They contain lime and are almost self-fluxing. The Warrior Basin coal is close which means that Alabama's steel industry has traditionally had low assembly costs for rawmaterials. However, the Birmingham district now also imports its ore by way of river barges from Mobile.

Fig. 25 – For the first time in modern history, pig iron production had a faster rate of growth than steel production in the years after the Second World War, which created a brisk demand for the ore fields. Both the Soviet Union and Japan passed the United States with their output of pig iron.

The modern oxygen processes have a good economy with a high share of pig iron in the charge, while the old open hearth furnace often had a charge of 50:50 for pig iron and scrap.

The many mini-steelworks, with their electric furnaces, create demand for scrap. For most metals, secondary recovery provides new metal with smaller inputs of energy than primary production.

Iron Ore 113

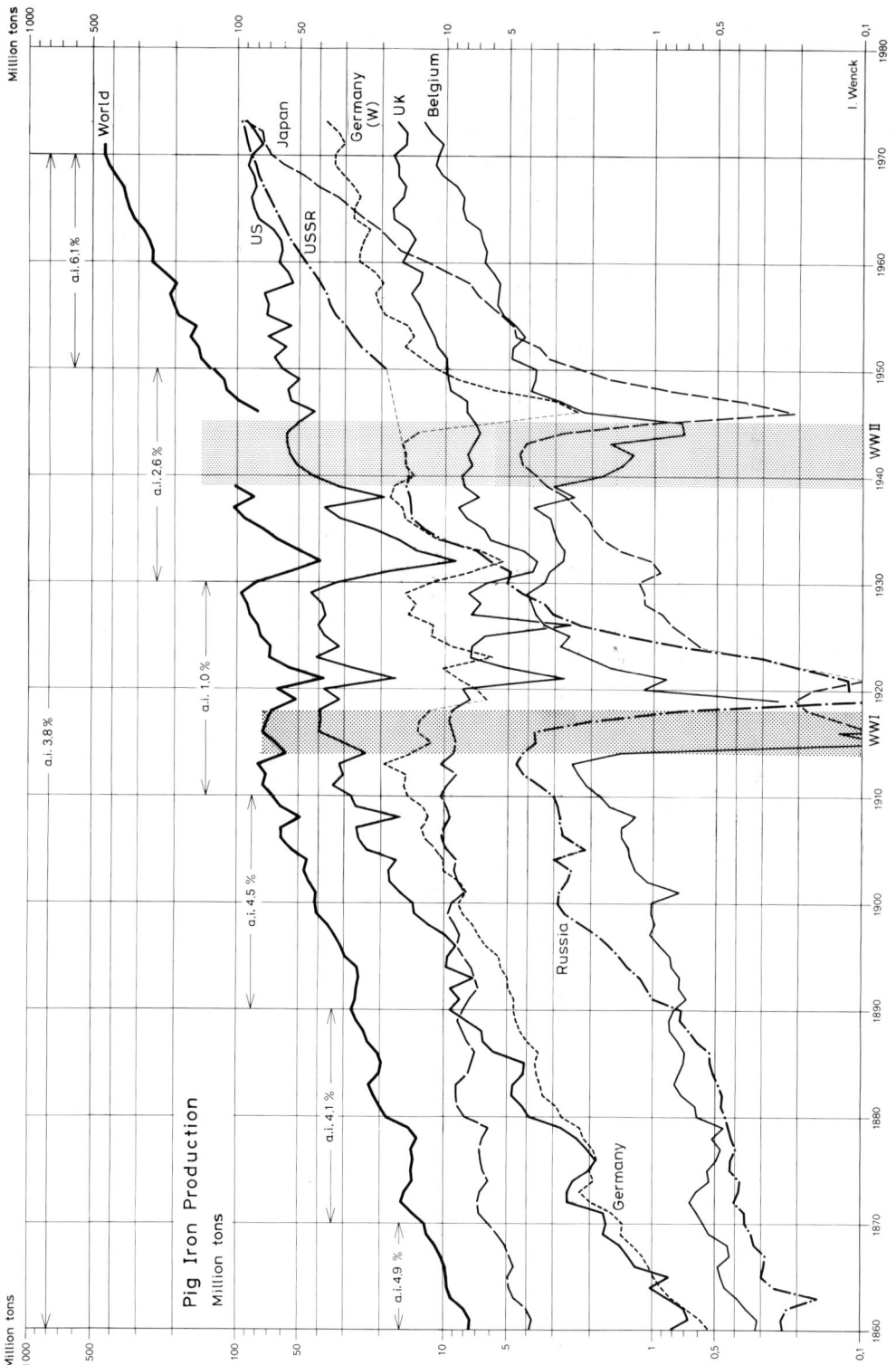

The open-pit at Daingerfield in northeast Texas averages less than 25% iron. The ore, limonite with some siderite, is beneficiated to 50% before being used. Western United States has three iron ore deposits, the open cuts at Cedar City in southwestern Utah being the most important. They contain magnetite of some 50% iron. The same holds for Eagle Mountain north of Desert Center in southern California. The Sunrise hematite deposits in Platte county, Wyoming, supply the blast furnaces of Pueblo, Colorado.

Canada — The Canadian deposits occur in four widely separated areas. The oldest mine was Wabana on Bell Island, Newfoundland (hematite, 53%), which was closed in the 1960s. It extends for a long distance under the ocean. This mine became Canadian when Newfoundland in 1949 became a province in the federation. The Wabana ore has a high silica content. Most of it was exported to Europe.

Ontario has three iron mines: at Steep Rock northwest of Lake Superior where a 40 m deep lake first had to be drained before mining could begin, first by open-pit and later by underground mining. The mine at Michipicoten is near Lake Superior. The Marmora mine east of Toronto contains taconite that is being beneficiated to 65% pellets.

The vast deposits of the Labrador Iron Trough, straddling the Quebec-Labrador border, rank internationally among the large postwar iron ore developments. The Trough contains high-grade ore that has been naturally beneficiated by weathering and low-grade siliceous ore that lends itself to economic beneficiation. The 575 km railroad from the high-grade deposit at Schefferville (Knob Lake) to Sept Iles (Seven Islands) was completed in 1954. Other ores in the area have also been developed. Wabush Mines and Carol Lake are served by Labrador City on a branch line of the mentioned railway. The Gagnon district (Mount Reed, Mount Wright) are on a separate railroad some 240 km north of Port Cartier in Shelter Bay. In these two districts low grade iron ore is beneficiated to pellets of 65% iron content. Early plans for ore developments near the western shore of Ungava Bay, with ore being stored on the west coast of Greenland for all-year shipments, have been shelved. The development of the Labrador iron ore strongly influenced the decision to deepen the St. Lawrence Seaway (1954–59) but the subsequent increase in the size of ocean-going ore carriers made flows through tidewater ports (Philadelphia, Baltimore) competitive with hauls through the Seaway, which has not handled as much iron ore as originally projected.

The least important Canadian ore district is in British Columbia (Quinsam Lake, Texada Island); the ore is shipped to Japan.

Latin America

Latin America has several large iron ore deposits. Some, e. g. in Brazil, have been known for a long time, but large-scale production is a post WW II phenomenon. In addition to Brazil, Chile, Columbia, Cuba, Peru and Venezuela are richly endowed.

Brazil – The Itabira ore and other Brazilian deposits near Belo Horizonte in Minas Gerais were known long before the First World War and projects were under discussion, but exploitation was not started until the 1940s when the Volta Redonda steel works was built and a railroad constructed to the export terminal at Vitória. In the early 1960s a new ore port was opened at Ponta do Tubarão. The state owned mining company, Cia Vale do Rio Doce, operates open cuts in the high-grade deposits (over 60% iron content) found in the lower grade siliceous ore, locally known as itabirite. Rapid expansion of the domestic steel industry and of ore exports to Western Europe, the United States, Japan and Eastern Europe made Brazil one of the fastest growing ore exporters in the 1960s.

Venezuela – In the 1940s the two leading American steel corporations, US Steel and Bethlehem Steel, developed ore deposits found in Venezuela on both sides of the Caroní River, a southern tributary of the Orinoco River. In the early 1950s Bethlehem started ore exports from El Pao east of the river and US Steel from Cerro Bolivar west of the river. Venezuela has built a heavy industrial complex at the confluence of the Caroní and Orinoco Rivers with a hydroelectric power plant, steel works, aluminum refinery etc. in addition to the ore shipping facilities (Puerto Ordaz). The new city, Ciudad Guayana, planned for a population of several hundred thousand people, is one of the largest new city projects in the world.

In contrast to most other new cities in the Third World it is not primarily planned for administration and other service functions but for heavy industry. The city is an important symbol for Venezuela's ambition to plow back the revenue from oil and iron ore into industry and infrastructure, to create an east-west regional balance, and to counter the economic dominance of the capital and the central region. The iron mines of Venezuela were nationalized in 1975.

Chile – Most Chilean mines are near ocean ports. Bethlehem Steel has operated its El Tofo and El Romeral open-pits since before the First World War. In anticipation of the new Panama Canal, opened in 1914, Bethlehem wrote a contract with the Broström group of Göteborg about hauls of iron ore between Chile and Sparrows Point, Baltimore, the first long-distance ocean shipments of ore with specially designed ore carriers. Because of the war actual shipments were not started until the early 1920s.

Peru – In comparison with Chile Peru is a newcomer among iron ore exporters. The first ore was shipped in 1953. It is produced in two strip mines in the southern part of the coastal desert. The Marcona mine, some 40 km inland from the port of San Juan, was nationalized in 1975. All facilities, including pelletizing plant (the ore is beneficiated to 70% before shipment), belt conveyor (between mine and coastal plants), and seawater desalting facilities, were built by an American mining company, owned jointly by Utah Construction and Mining and Cyprus Mines. The Acari mine is located somewhat further inland. As in Chile, some ore is used by the domestic steel industry.

Cuba and the Dominican Republic have some production of iron ore. Reserves are large in Cuba but the interwar exports to the United States were discontinued after the Second World War when richer ores were developed elsewhere.

Western Europe

In Western Europe, France and Sweden are leading producers but also England, Spain, Norway, Germany (W) and Finland have iron ore mines.

Sweden – For Sweden's position as the leading exporter of iron to the world market in the early 18th century and for its modern position as a major producer of special steels the many deposits of exceptionally pure ores in the Bergslagen district northwest of Stockholm played a major part. The most famous is Dannemora. However, Sweden's export of iron ore is a more recent development that goes back to the invention (1878) of the basic Bessemer steelmaking process (Thomas-Gilchrist process) which permits ore with a high phosphorous content to be used.

The introduction of Thomas converters on the continent created a market for the large deposits of high-grade but phosphorous magnetite ore at Kiruna and Gällivare in Lapland and Grängesberg in Bergslagen with Narvik, Luleå and Oxelösund as ore shipping ports. About half the ore from these mines have a low phosphorous content. The Kiruna ore was found in two mountains, Luossavaara and Kirunavaara, which gave their names to the mining company, LKAB, that operates all iron mines in Lapland. In the late 1950s the Swedish government took almost full control of the company, using its option to buy the shares of the private Gränges group. This group before 1957 was the operator in the Lapland mines and the state only a financial partner. LKAB and Gränges have a common sales organization, AB Malmexport, which also sells the Nimba ore of Liberia.

Ore production at Kiruna is dominated by two mines, the Kirunavaara mine which is now exploited underground with tunnel access and the Svappavaara mine, 40 km southeast of Kiruna. The Gällivare district has about one-fourth the production of Kiruna (45 and 12 million tons). Grängesberg is much smaller (4 million). All other mines combined produced 7 million tons of ore in 1974.

Norway since 1906 exploits a low-grade iron ore deposit in the extreme north at Kirkenes. The ore is beneficiated to over 2 million tons of pellets. Small iron mines are found near the steelworks at Mo i Rana and at Malm on the Trondheim Fjord. Norway is a small iron ore exporter in its own right; in addition it has the large ore port of Narvik, serving the Kiruna mines.

Finland is an even smaller producer of iron ore with mines on Jussarö in the Ekenäs Archipelago in the south and several small mines in the north, including Otanmäki south of Oulujärvi, which is a magnetite-ilmenite deposit, Raajärvi on the Kemi River, and Kolari on the Torne River. In 1972 the Finns signed a contract with the Russians about developing a large iron ore deposit on the Russian side of the border at Kostamus on the latitude of Raahe. The Finns will be paid in ore deliveries for the Rautaruukki steelworks at Raahe. Sales to the Swedish steelworks at Luleå have been mentioned. This would imply an ore railroad almost from Moscow (Cherepovets) to Narvik.

France – Lorraine dominates the French iron ore production. The minette ores are oolithic limonites with low iron content. The Thomas process turned these phosphorous ores into a remarkable asset. The ore field straddled the Franco-German border of the day and iron works were built on both sides. After WW I this heavy industrial district, centered on Nancy and Metz, was within France. The low iron content and difficulties in beneficiating the ore makes it imperative to locate the blast furnaces on the ore field. Lorraine was a showpiece for Jean Monnet and his vision: If the heavy industries of Western Europe were integrated, wars would become impossible. The European Coal and Steel Community (1952) was first of the communities created to achieve this goal.

The iron ores of Luxembourg are a continuation of the French fields in Lorraine. The iron and steel industry based on the ore dominates the economy of this small country.

Germany, whose declining domestic output accounts for only a very tiny share of the ore consumption, has had the major field at Salzgitter near Braunschweig, which was developed to supply the large steel mill built in the 1930s. In 1976 the Elbe-Seiten-Kanal was opened. It was expected to carry some 5 million tons of

118 Metals and Other Minerals

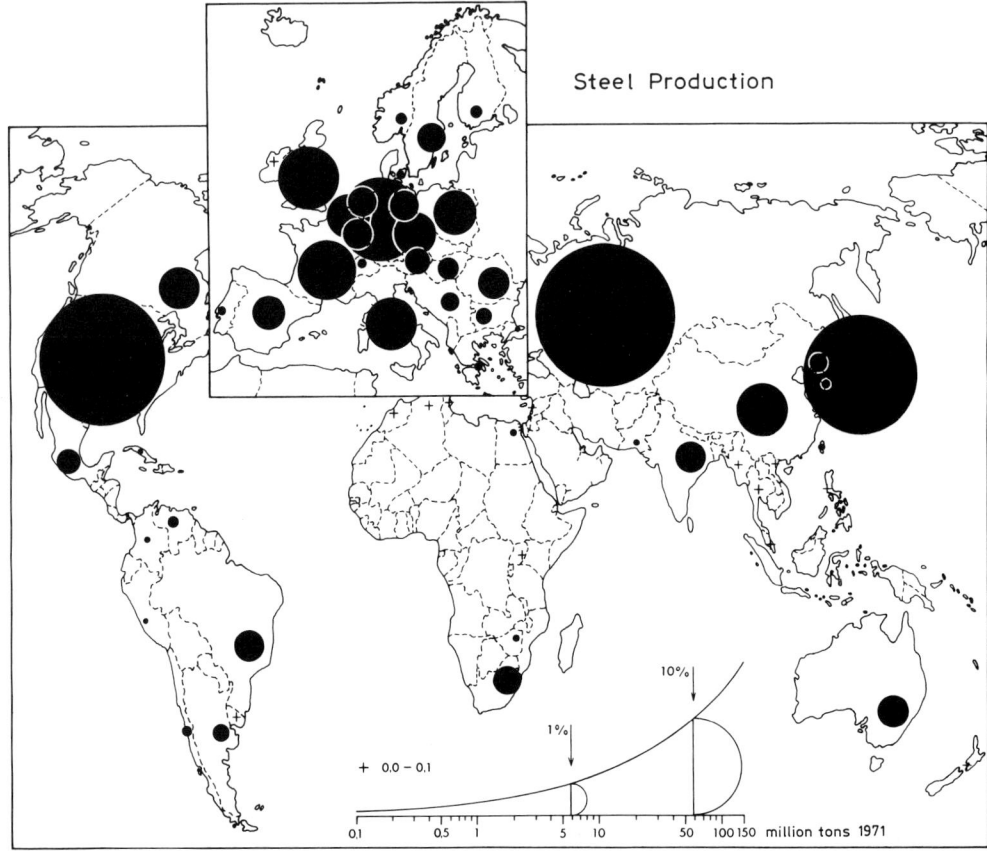

Fig. 26 – United States, the world's leading steel producer since the end of the 1880s, was surpassed by the Soviet Union some years in the early 1970s. Before the mid-1970s, Japan would have surpassed both superpowers if the recession had not forced some revisions of plans. The position of Western Europe has declined but EEC-9 still surpasses all countries and EEC-6 has a larger production than Japan.

imported ore from Hamburg to Salzgitter. The railroad, however, has succeeded in attracting this transport.

The ores at the rivers Sieg and Lahn south of the Ruhr area supported a local iron industry long before the modern developments on the Ruhr got under way about 1850. Mining in this district expired after WW II.

United Kingdom – The British iron ore fields supply $1/3$ of the domestic consumption expressed in iron content as against $2/3$ in the late 1930s. Almost all domestic ore comes from low-grade (28%) Jurassic sedimentary beds, similar to those of Lorraine and Salzgitter, which are mined by opencuts in an area stretching from south of Lincoln to north of Oxford. As on the continent the

Tab. 23 — International Trade in Iron Ore 1971 (SITC 281)
(million US $)

Imports		Exports	
Japan	1 332	Australia	453
Germany (W)	483	Canada	409
United States	453	Sweden	248
United Kingdom	263	Brazil	237
Belgium	196	Venezuela	166
Italy	133	Liberia	161
France	108	India	137
Netherlands	57	Mauritania	75
Total (excl CC)	3 178	Total (excl CC)	2 337

CC = Communist countries.
Source: YITS 1972–73.

Tab. 24 — International Trade in Pig Iron 1970 (SITC 671)
(million US $)

Imports		Exports	
Japan	246	France	112
Germany (W)	148	Norway	109
Italy	121	Germany (W)	103
United Kingdom	117	United States	82
France	80	South Africa	67
United States	78	Canada	49
Sweden	73	India	49
Belgium	49	Sweden	34
Total (excl CC)	1 135	Total (excl CC)	852

CC = Communist countries.
Source: YITS 1972–73.

ore is processed near the ore fields. The Cleveland ores, that used to feed the local iron and steel industry of Middlesbrough, are now exhausted. This steel district, as well als those of South Wales and Scotland, depends on imported ore.

Austria has an expanding production of low-grade ore at Erzberg (siderite) and at Hüttenberg in Kärnten. The Austrian steel industry, with mills at Linz and Donawitz, contributed one of the major postwar innovations in steel making, the LD-process, which has greatly changed the structure of the world's steel industry after 1952.

120 Metals and Other Minerals

Spain is an iron ore producer of long standing; in the 17th and 18th century it was a major exporter of iron and in the early decades of the ore boom after the introduction of modern steel processes in Western Europe (from about 1860) it was a leading exporter of iron ore. The peak was reached around the turn of the century with some 10 million tons a year. Of the present output only marginal quantities are exported, primarily to Germany and South Wales. The ore fields at Bilbao and Santander have attracted a domestic iron and steel industry.

Central Europe

Yugoslavia has iron ore fields north of Sarajevo and between Zagreb and Banjaluka. The Comecon countries of Central Europe have only small domestic production of low-grade iron ore (Romania, Bulgaria, Poland, Czechoslovakia, Hungary and East Germany), each ranging between 1.0 and 0.1 million tons iron content. They import ore from the Soviet Union by rail or by river boats. Krivoj Rog is a major supplier of ore to the steel industry of the Comecon countries.

USSR

The Soviet Union in 1958 surpassed the United States as an iron ore producer and in 1970 accounted for $^1/_4$ of world production.

The iron industry had its origin near Tula and Lipetsk south of Moscow in the 17th century. It was based on local iron ore and charcoal and produced gun barrels and rifle bores but also saithes, sickles and nails for a primitive market. To win his war against Sweden Peter the Great initiated an even larger industry in the Urals. This region with its vast forests, waterfalls and iron mines in the second half of the 18th century was the foremost metallurgical area in the world. The Urals in 1767 produced 56000 tons of pig iron or twice as much as England. Russia at this time dominated the world market for bar iron. The Urals still is an important center for the Russian special steel industry which long was based on charcoal iron.

The Ukraine — The Industrial Revolution came to Russia with a delay of two or three decades compared to Western Europe and eastern United States. Construction of iron mills to utilize the coal of the Donets Basin and the ore of Krivoj Rog was started in the 1860s; rapid expansion in the Ukraine came after 1880 with the inflow of French and Belgian capital and technology and with the construction of railroads. By 1900 the Ukraine had replaced the Urals as the foremost metallurgical region of Russia and in 1913 the Ukraine produced 75% of the iron ore of the country, 69% of its pig iron and 57% of its steel.

Fig. 27 — The steel production of the world had its fastest growth before the First World War. The rate of increase for the 1865–1970-period was 7.2% a year or a doubling every ten years.

Iron Ore 121

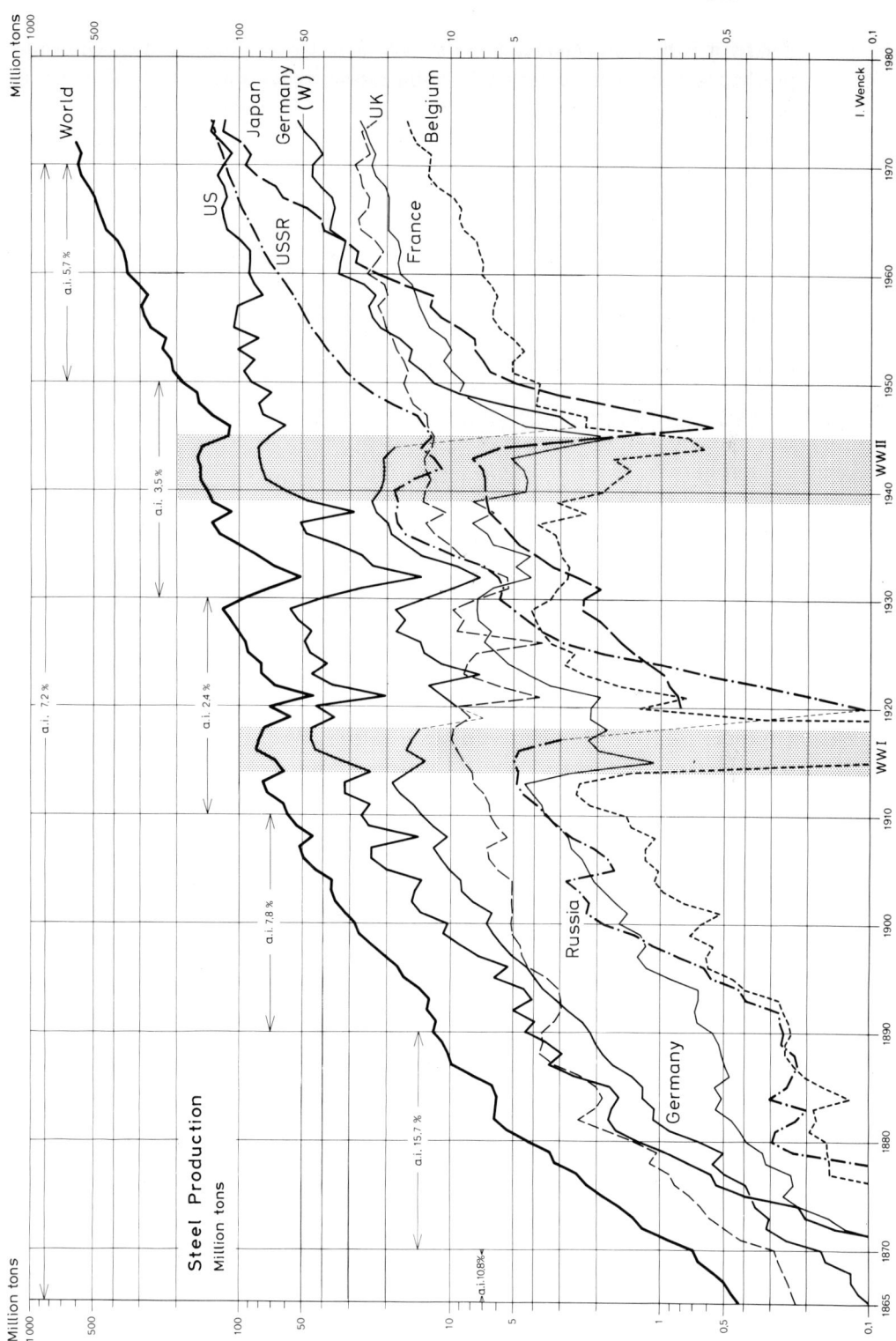

Krivoj Rog is still the leading iron ore field in the Soviet Union. Until the postwar period only the rich hematite ore, (57%) was mined. The ore is obtained in shaft mines of down to 800 m. More recently the low-grade quartzite-hematite ore (36%) has been worked. Being close to the surface it can be strip mined. In the beginning of the 1970s opencuts accounted for 30% of the metal. Production at Krivoj Rog has followed a steeply rising curve:

1913	1928	1940	1950	1958	1970(p)	
6	5	18	21	47	83	million tons (p-plan)

Two more mining districts in the Ukraine are important. The new field at Kremenchug has the same type of ore as Krivoj Rog and in addition an excellent location with regard to the steelworks. The oolithic limonite ore (37%) on the Kerch Peninsula were mined already before the First World War. Production has increased strongly since the 1950s and in 1970 accounted for some 20 million tons. Some of the ore is processed in local blast furnaces.

Kursk Magnetic Anomaly (KMA) – A vast iron ore deposit, long considered to be the largest in the world, with a central location in relation to Tula, Lipetsk and Donbass, was made available in the late 1950s, when the Soviets had solved the technical problems of the water-logged overburden. The ore in the KMA underlies an area of $120 000$ km^2, mostly at a depth of 120–300 m, and is estimated to hold 200 billion tons. Most of the ore mass, which holds 43% iron, also presents metallurgical difficulties because of the high silica content. However, in 1953 ores with an iron content of 60% and relatively low in silica were discovered. KMA after that has been expoited on a relatively large scale. In 1972, the Soviets published plans for expanding Lipetsk to become the world's largest steel mill (14 million tons), surpassing Magnitogorsk (12 million tons). Lipetsk is strategically located in relation to the steel markets and the KMA ore. Iron ore mines have been opened at Stary Oskol and Mikhailov.

The Urals – A new expansion period in the Urals came with the second 5-year plan 1932 and the exploitation of the magnetite ore (50%) in Mount Magnitnaya at Magnitogorsk and the establishment of the Urals-Kuzbass Combine (UKC). The marriage of the coking coal of the Kuznetsk area in Siberia with the iron ore in the Urals became one of the first great showpieces of Soviet central planning, after the reconstruction period.

The old time of glory of the Urals as an iron district had been based on charcoal; the new steel industry was to build on coke. In the old time, the southern part of the highland area had less attraction because of its lack of forests. For a cokebased iron industry, the whole Urals region had the disadvantage of a great shortage of coal of coking quality. By building steel mills in the Urals and in Kuzbass, at both ends of the long raw material shuttle, freight cars could be utilized in both directions, coal moving westward and iron ore eastward. The relocation of steel capacity during World War II further strengthened the position of the Urals; the German occupation of the Ukraine proved the UKC to be vital to the survival of the country. It gave the depth in the economic defence of the Soviet Union that later made possible a rolling back of the enemy's positions. The Urals were strengthened as a steel district during the war.

Scattered fields – A large number of smaller mines have long been operating within the vast territory of the Soviet Union. Lipetsk and Tula in the center serve the local steel industry. The Rustavi steel mill in Georgia processes iron ore from Dashkesan in Azerbaijan. The coal-field oriented Karaganda mill gets some of its ore from the 'local' Atasu field and the rest from Magnitogorsk. The large Cherepovets mill, opened in 1955, has long rail hauls for its coking coal from Vorkuta and its iron ore from Olenegorsk in the Kola Peninsula; in addition it is about 500 km from its main market (Leningrad). Cherepovets will get a new ore base with the completion of the Kostomukscha (Fi Kostamus) mine in 1977. Some 24 million tons of ore will be beneficiated to 8 million tons of pellets. The mine is 65 km west of Juskosero, the present terminal of the railroad, 30 km from the Finnish border and 1 120 km from Cherepovets. The ore from the Kola Peninsula, a further 320 km to the north, will probably be exported through Murmansk.

Asia

Japan – In Asia, Japan is the dominant steel producer with an output that is rapidly approaching those of the Soviet Union and the United States. But Japan, the most expansive of the major steel producing nations, has hardly any domestic ore resources; Japan is the leading market in the world for iron ore as for many other raw materials.

China on the other hand, in spite of considerable expansion after 1949, is still a relatively minor steel producer (some 20 million tons in 1972) but has large iron ore reserves. The leading iron ore mine is at Anshan in former Manchuria which was exploited by the Japanese in the interwar period. Other major mines are in the Yangtze Valley and in Hupei.

India has long been known for its large iron ore reserves, most of which are located in the states of Bihar and Orissa. The Tata Works at Jamshedpur, established before the First World War, are near the ore fields, some 100 km south of the Raniganj coal field and 200 km west of Calcutta. In addition to the large, purely Indian Tata Works, India since independence has built several

Tab. 25 – International Trade in Steel and Semimanufactured Steel Products 1970
(SITC 672–679)
(million US $)

Imports		Exports	
United States	1 952	Japan	2 837
Germany (W)	1 526	Germany (W)	2 427
France	1 224	Belgium	1 927
Italy	850	France	1 444
Netherlands	680	United States	1 294
Switzerland	433	United Kingdom	810
Belgium	420	Sweden	569
Canada	418	Netherlands	477
United Kingdom	417	Italy	443
Spain	339	Canada	363
Total (excl CC)	13 458	Total (excl CC)	13 697

CC = Communist countries.
Source: YITS 1972–73.

integrated steelworks with foreign assistance. The iron ore mines not only supply the domestic steel industry which in the 1960s and early 70s remained stagnant around 6 million tons but also account for growing exports. In the early 1970s India was on a par with China as an ore producer, each exceeding 20 million tons in iron content. Goa, a Portuguese colony for 451 years until 1961, when it was incorporated into India, had considerable exports of iron ore in its own right.

Other Asia – Other exporters of iron ore in Asia have been Malaya and the Philippines, both with Japan as the major customer. High-grade hematite has been produced in several mines in Malaya (in Trengganu, Johore and Perak) before and after the Second World War, but in the early 1970s the Japanese mining interests almost discontinued a production that peaked at over 6 million tons in the early 1960s. In the Philippines high-grade hematite ore is mined at Balucan in Central Luzon and at Camarines Norte. The vast lateritic deposits (48% iron) at Surigao in northern Mindanao are of great potential importance. But Japan, in the 1950s anxious to replace its prewar supplies from Manchuria and Korea, lost some of its earlier keenness on the development of secondary iron ore fields in Asia, when the vastness of the Australian iron ore deposits became known.

Australia

Australia in the 1930s was seen as a continent poor in iron ore. Broken Hill Proprietary Co. (BHP), which from 1935 formed a monopoly in the iron and steel industry of the country with dominating steel centers in the coal ports at Newcastle and Port Kembla (Wollongong), had its ore base in Iron Knob, South Australia, with loading terminal at Whyalla. When Western Australia in 1935 became interested in exporting iron ore to Japan and in developping the ore at Yampi Sound, known for a long time, it met with resistance. The federal government in 1938 enacted an export embargo. This may partly have been for geopolitical reasons but the smallness of Australian reserves was a major argument. The policy was not changed until 1960/61 when low-grade ore had been found at several places.

As soon as the export embargo had been abandoned the owner of some sheep stations in northern Western Australia, Lang Hancock, got in touch with the Conzinc Riotinto group and informed them that in 1952, when searching for manganese ore from an airplane, he had found large deposits of iron ore. This started the great iron ore rush in the Pilbara district which rapidly carried Australia to the first place among the iron ore exporting countries of the world. Among the mines are Mt. Newman, Mt. Whaleback, Mt. Goldsworthy, Mt. Tom Price, and Mt. Enid. The ore is shipped through three ports, Port Hedland, Port Lambert, and Dampier. All could have been sent through one port, but the Japanese buyers insisted on separate ports. They did not want to risk the Japanese steel industry coming to a standstill if a single ore terminal was hit by a natural calamity of the sort that hit Darwin in 1974.

Africa

In Africa, the Maghreb countries and South Africa have long produced iron ore, but it is only after the Second World War that tropical Africa has become a major supplier of ore to the world market. Algeria, Morocco and Tunisia are small producers by modern standards; South Africa produces more than twice as much as the three countries combined.

South Africa – South African reserves are very large, but some are in the form of titaniferous magnetite which presents problems for ordinary blast furnaces. The Krupp-Renn process is successfully used for reduction of this ore separating the iron (as sponge iron) from the titanium oxide slag. The largest mines are in high-grade hematite deposits of the Lake Superior type near Postmasburg in Griqualand West and at Thabazimbi north of Rustenburg in the Transvaal. Both iron mines are in remote semi-arid areas. In the mid-1960s Swaziland got an iron mine that produces more ore than the two small mines in Southern

Rhodesia combined. South Africa has also become involved in the Japanese ore trade. The development at Sishen was made possible by Japanese long term contracts and a state guarantee which facilitated loans in the eurodollar market of London. Saldanha on the Atlantic coast will be the terminal of a 800 km railway and will also get a steel works.

Liberia – The largest producer in tropical Africa is Liberia followed by Mauretania and Angola. Liberia has several large iron ore projects of which the joint Swedish-American-Liberian Lamco venture at Nimba is the most important. It is served by the ore port of Buchanan. Three mines, Bong, Bomi Hills and Mano River are served by Monrovia.

Mauritania – The rich ore from Zouerate (Fort Gouraud) in Mauritania is shipped 650 km by rail to Nouadhibou (Port Etienne).

Central African Republic – Analysis of material obtained from an American satellite has revealed an exceptional magnetic anomaly near the Equator in Africa, The Bangui Magnetic Anomaly. Iron and uranium ore have been found on the ground also, but the exact extent of this vast deposit is by no means known. Remote sensing techniques have also been used for finding other ores, especially in out-of-the-way areas of the world.

Tab. 26 – World Production of Major Metals
         (million tons)

Metals	Early year	Production			Longterm growth, accum, p c p y	Modern growth, accum, p c p y
		Early year	1953	1970		
Pig iron	1867	10.1	168.2	440.5	3.7	5.8
Aluminium	1890	0.00018	2.5	9.7	14.6	8.3
Copper	1879	0.15	2.8	6.3	4.2	4.9
Lead	1882	0.41	1.9	3.3	2.4	3.3
Zinc	1873	0.14	2.3	4.8	3.7	4.4
Tin	1867	0.025	0.18	0.18	1.9	0.07

accum = accumulated, p c p y = per cent per year
Longterm = Early year–1970, Modern = 1953–1970

Alloy metals

All steels in addition to iron and carbon contain small but important amounts of other elements, e. g. silicon and manganese. To give ordinary carbon steel characteristics, which are in demand among constructors one or more metals are added, as a rule in small quantities, often less than 1%. The material added to the charge of the steel furnace is usually manufactured at special ferroalloy works, which are heavy consumers of electricity and therefore as a rule located near hydroelectric power plants (e. g. in Norway or Canada). The ferroalloys produced in these works contain a high percentage, some 40–90%, of the 'other' element (chromium, cobalt, manganese, molybdenum, nickel, tungsten, vanadium, silicon or cerium). Some elements are added as pure metal, e. g. aluminium, titanium, zirconium or nickel. The listed elements as a rule have other uses as well.

The term alloy metal usually refers to those metals added to steel. However, also non-ferrous metals are alloyed. Common alloys of the latter type are bronze (copper and tin) and brass (copper and zinc). The alloy metals are usually mined only in a few countries. They are indispensable in the manufacture of armaments and are often referred to as strategic materials. Albert Speer, Hitler's minister of war production, in his memoirs underlines that German stocks of chromium, silicon, molybdenum, nickel, wolfram, and manganese were very short during the Second World War.

World production of manganese and chromium ores amount to millions of tons but for the other alloy metals quantities are rather small.

Manganese

Manganese (Mn) in small amounts is used in the production of most kinds of steel. Normally the manganese content is about 1%. The purpose of the input is to neutralize the negative effect of sulphur on the high-temperature strength of steel. Larger amounts, up to 14%, are added to give steel a hard surface but tough core, needed in e. g. rock crushers and railroad switches. Manganese is also used in the manufacture of bricks, glazed pottery, plastics, floor tile, glass, varnish, and dry-cell batteries.

Before the Second World War, the Soviet Union and India were the dominating producers. Nikopol in the Ukraine and Chiatura in Georgia are the leading Soviet mines. Other deposits are at Tokmak in the Ukraine, Ulu-Teljak in the Urals, Usa in Kuzbass and Jashno-Chingan near the Amur River. India has several open pit mines in the center and in Goa but grades are poor and

India's relative position has declined. China, on the other hand, has had a rapid postwar expansion and is now almost on a par with India.

After the war Africa has emerged as a major producing region: Postmasburg and Kuruman west of Kimberley are centers of manganese mining in South Africa. Mouanda near Franceville in the interior of Gabon has one of the major manganese reserves in the world. It has been mined since 1959 by US Steel and French interests. The ore is shipped by way of Pointe Noire. Nsuta in Ghana, north of Takoradi, has a large open pit, in operation since 1916. The production of Zaire started in 1948 at Kisenga and Kasekelesa in south western Shaba (Katanga) near the Benguela railroad.

Also Brazil has become a major producer. Deposits of manganese ore occur near Macapá in Amapa north of the lower Amazon River, in the plateau of Minas Gerais, and at Urucum near Corumba in western Mato Grosso. Small manganese deposits occur in several countries of South, Central and North America.

The conspicuously small American production of manganese may help explain the interest shown by the United States in exploiting the tremendous deposits of manganese nodules on the deep sea floor, especially in the Pacific Ocean. The nodules also contain other metals, copper, nickel, and cobalt, in amounts that may cover the needs of mankind for thousands of years. It has been stated that minerals are formed here faster than they are being used on Earth.

The salvage of a Soviet submarine that sunk in 1968 at a depth of 5180 m some 1200 km northwest of Oahu, Hawaii, shows that the United States in 'Glomar Explorer' has the world's most advanced technology for mineral exploitation on the ocean floor. The ship, built by Howard Hughes, on a contract with the CIA, is 188 m long with a 35 m beam and has a loading capacity of 36000 dwt tons.

If this type of mineral exploitation can be made competitive remains to be seen. Mineral exploitation was given as a pretext during the salvage operations in 1974. The submarine sunk in an area with a strong concentration of manganese nodules.

Tab. 27 — World Production of Manganese Ore 1970
(thousand metric tons)

USSR	6984	India	1651	Australia	804
South Africa	2679	Gabon	1453	Ghana	405
Brazil	1928	China	998	Zaire	346
World	18493				

Source: MY 1970, Vol. 1

Chromium

Chromium (Cr) is second only to manganese among the alloy metals. Small amounts increase the hardenability of steel, especially if nickel is also present. Steel with 1.2% Cr + 4% Ni is used in tools. Larger amounts, 12–15%, increase high temperature strength and corrosion resistance as well as resistance to wear (kitchen utensils, cutlery, oil burner components, bearings). Stainless steel (Ni 8%, Cr 18%) is used widely in machinery where steam, water, moist air, or acids would corrode ordinary steels quickly. Chromite ore is also used as a refractory material in basic brick linings of various metallurgical furnaces (chromite or chrome-magnesite brick). Chromite (FeO. Cr_2O_3) is the only chrome mineral of significance; it is the raw material of ferrochrome.

Five countries dominate the output of chromite ore. The Soviet union has mines at Sarany north of Sverdlovsk, and at Chromtay in the southernmost Urals. Africa has an even larger production with mines at Rustenburg in South Africa and at Selukwe and Kildonan in Rhodesia. In Asia production is dominated by the Philippines with the largest mine at Zambales in northern Luzon, and by Turkey. Fethiye in the southwest and Guleman in the east are the leading Turkish mines. The Balkan countries of Europe have some mines of which Kukes in Albania is the largest. The Western Hemisphere has very little chromite ore.

Tab. 28 – World Production of Chromite 1970
(thousand tons)

USSR	1750	Turkey	477	India	266
South Africa	1427	Albania	454	Madagascar	141
Philippines	576	Rhodesia	363	Finland	121
World	5920				

Source: MY 1970, Vol. 1

Nickel

Nickel (Ni) gives strength to iron and other base metals. For heat and corrosion uses, nickel has no substitute. In jet engines, where some parts operate at white heat, in gas turbines and rockets nickel is used. It is used with chromium in the making of stainless steel. Among the nickel-copper-alloys is monel metal (67% Ni, 28% Cu, 5% Fe) which is highly corrosion resistant to salt water and therefore is employed in pumps, propellers and mine screens. Permalloy (80% Ni, 20% Fe) is the trademark of an easily magnetized and demagnetized alloy used in electromagnets. Alnico, an Al-, Ni-, Co-, Fe-alloy, is a trademark for strong permanent magnets.

Tab. 29 – World Production of Nickel 1970
(thousand tons)

Canada	277	New Caledonia	105	Australia	28
USSR	110	Cuba	35	Indonesia	17

World	621				

Source: MY 1970, Vol. 1.

The mining of nickel is more localized than that of most other metals. The complex nickel ores usually also contain copper and other metals. They are associated with basic igneous rocks. Some 80% of production has been obtained from underground sulfide ores, but some 80% of known nickel reserves are in laterite deposits, which have proved difficult to process. Since 1900 Canada has supplied most of the world's nickel with Sudbury by far the leading nickel-area. One mine here is more than 2000 m deep. Newer mines are operated at Lynn Lake and Thompson in northern Manitoba. Several companies participated in the development of the large nickel deposits at Sudbury, which were discovered in 1883 in connection with the construction of the Canadian Pacific Railway. The companies were merged in 1902 and formed the International Nickel Company (Inco). German born Dr. Ludwig Mond developed a chemical process in England for the recovery of nickel from sulfide ores. Mond Nickel in 1929 was acquired by Inco. A third company with a third nickel recovery method, electrolysis, was established in 1929 at Sudbury (Falconbridge Nickel). This company has an electrolytic refinery at Kristiansand on the south coast of Norway.

In the Soviet Union nickel is mined at Sverdlovsk and Orsk in the Urals, in the Kola Peninsula at the former Finnish Petsamo[19], now Pechenga, and at Norilsk in Siberia.

The third important nickel producer is New Caledonia where a lateritic nickel-bearing mineral is exposed over much of the island. These deposits were discovered in 1854 and developed by the Rotschild-company La Société Le Nickel (SLN), which dominated the world market until the turn of the century. The French armament factory Schneider at Le Creusot became known for its nickel steel.

Similar but smaller deposits occur in the Oriente Province of Cuba, which were developed during World War II by the American Freeport Sulphur Co, now Freeport Minerals, which is also engaged in the mining of oxide ores elsewhere. This low-grade ore has proved costly to exploit. Nickel oxide is obtained which is used directly in the steel charge. Also Brazil, Venezuela, Burma, Indonesia (Sulawesi/Celebes), Australia (Queensland) and other tropical areas have such deposits but exploitation has been slow to start. Australia in addition found deposits of sulfide ore in the 1960s in and near the old gold district at Kalgoorlie in West Australia.

Tab. 30 — International Trade in Nickel in 1970 (SITC 683)
(million US dollars)

Imports		Exports	
United States	317	Canada	446
Germany (W)	180	United Kingdom	143
United Kingdom	129	Norway	106
Sweden	80	New Caledonia	73
France	64	United States	67
Japan	60	Germany (W)	53
Italy	59	France	36
World (excl CC)	1 097	World (excl CC)	1 015

CC = Communist countries
Source: YITS 1972–73.

Tungsten

Tungsten (W, wolfram; Swedish: tung sten, heavy stone) is used in steels for high-speed cutting tools (18% W) and in hard metal for rock drills as well as in pure form in electric-lamp filaments. It has the second highest melting point among the elements.

Hard metal or cemented carbide is a sintered mixture of tungsten carbide and cobalt in which cobalt is used as a binder. Tungsten carbide (WC) is almost as hard as diamond. It is used in cutting tools and abrasive wheels. Steelite, a W-, Co-, Cr-alloy, is used for hard facing of other metals: armor plates, guns, and armor-piercing projectiles.

The principal tungsten mineral is wolframite, $(Fe, Mn)WO_4$, which contains only 3% tungsten. It is concentrated to about 60% tungsten oxide before being shipped to the market. Another tungsten mineral is scheelite $CaWO_4$. China has long been the leading producer, primarily in Kiangsi and Kwantung provinces. The Soviet Union, the second largest tungsten producer, has many mines at Tyrnyauz in the Caucasus, Majchura north of Dushambe and Bogutin east of Alma Ata. In Korea production is evenly divided between the northern and southern parts.

Since Communist countries accounted for 60% of the world output in 1949 many countries in the West were combed to find new deposits. In the United States 700 mines were producing in 1956 with federal support; the last two were closed in 1959. Among mines that have produced tungsten off and on are Climax, Colorado, Mill City, Cherry Creek, and Ely, Nevada, Pine Creek near Bishop, California, and Townsville, North Carolina.

Tab. 31 — World Production of Tungsten 1970
(tons)

China	7973	North Korea	2138	Portugal	1783
USSR	6704	South Korea	2067	Canada	1339
United States	3672	Bolivia	1843	Australia	1243
World	33522				

Source: MY 1970, Vol. 1.

In South America some tungsten is produced in Peru, Bolivia, Brazil, and Argentina. The leading producer in Europe is Portugal with largest mines at Borralha and Panasqueira. Africa has small mines in South Africa, Rhodesia and Zaire.

Antimony

Antimony (Sb; Stibium) is found in stibnite, Sb_2S_3, the only important antimony mineral. Many copper, silver, gold and lead ores contain some antimony. The metal has a low melting point and expands during solidification. As much as 30% antimony is alloyed with lead in type metal to increase the hardness of lead. Lead-antimony-alloys are also used in pipe, electric cable coverings, buckshots, solder, foil, and storage battery plates. Antimony compounds enter paint pigments, fire proofing materials, glass, and rubber.

Tab. 32 — World Production of Antimony 1970
(tons)

South Africa	17000	USSR	6700	Thailand	2400
China	11800	Mexico	4500	Jugoslavia	2000
Bolivia	11500	Turkey	2800	Morocco	2000
World	66300				

Source: MY 1970, Vol. 1.

China has long been the leading producer. In large deposits in Hunan and Kwangsi provinces the ore is broken and sorted by hand; concentrated ore carries 60% antimony. South Africa, which now tops the production table, obtains its antimony as a by-product from a gold mine in the Transvaal. Bolivian antimony is mined with many other metals at Potosí and other places in the tin belt of the Eastern Cordillera. The Soviet Union has large mines at Razdolinsk in central Siberia, Turgay east of the Urals and Frunze in eastern Turkestan. The largest producer in Mexico is San Luis Potosí; several lead mines yield antimony as a by-product.

Molybdenum

Molybdenum (Mo) has a high melting point and adds strength and creep resistance to steel at high temperatures. At low temperatures molybdenum remarkably increases the corrosion resistance of steel. Molybdenum can be used as a substitute for tungsten, which was of importance during the Cold War, given the distribution patterns of the two metals. Molybdenum steel is used in propeller shafts, gun barrels, and high-speed tools. Molybdenite, MoS_2, is the only important mineral. The dominant deposit is Climax, Colorado, at over 4000 m altitude. The low-grade ore is crushed and concentrated through flotation. Tailings are processed in by-product plants for the recovery of tungsten and tin. The balance of the American molybdenum is obtained as a by-product, primarily from copper mines in Utah and Arizona. United States has often accounted for $3/4$ or more of the world's output. Canada has a series of copper-molybdenum and molybdenum ores in British Columbia.

The leading Soviet mine is at Tyrnyauz in the Caucasia. There are smaller mines elsewhere. Erdenet in Mongolia, an open pit 320 km northwest of Ulan Bator which is under development with Soviet assistance, is expected to yield several thousand tons a year when put into operation at the end of the 1970s. Chile obtains most of her output from the copper mines at El Teniente and Chuquicamata. China's leading mine is at Chin Ling in the central part of the country.

Tab. 33 – World Production of Molybdenum in 1970
(thousand tons)

United States	50	Chile	5	Japan	0.4
Canada	16	Peru	0.5	Norway	0.3
World	74				

Source: MY 1970, Vol. 1.

Vanadium

Vanadium (V) in small amounts increases the toughness, hardness and fatigue resistance of steel. Such steel is used in armor plate, rock drills, springs and other components in automobiles, locomotives, and guns.

In minor amounts vanadium is present in most rocks but concentrations rich enough to be exploited are relatively few. United States produces vanadium in the Colorado Plateau where the mineral carnotite, $K_2O\ 2UO_3\ V_2O_5\ 2H_2O$, was mined by private companies under a federal AEC contract for its uranium content yielding vanadium as a by-product. This came to a stop when the Atomic Energy Commission could obtain cheaper uranium from rich deposits in Zaire, Canada and other countries. In the Colorado Plateau vanadinite $Pb_5Cl(VO_4)_3$, is mined for both vanadium and lead. South Africa is now the dominant producer with the chief mine at Middleburg in the Transvaal. Otanmäki, Finland, and Berg Aukas in Namibia are important producers. Gora Kachkanar in the Urals is the leading Soviet deposit. Peru, once the world's leading producer, and Zambia have suspended production after depletion of their better ores.

Tab. 34 — World Production of Vanadium in 1970 (thousand tons)

South Africa	7 346	Finland	1 315	Namibia	363
United States	4 824	Norway	980	France	91
World	14 918				

Source: MY 1970, Vol. 1.

Cobalt

Cobalt (Co) in steel, especially if it also contains tungsten, increases its hardness and high-temperature strength. Cobalt alloys are used for cutting tools in lathes (high-speed steel and hard metal) and for components in jet engines and atomic

Tab. 35 — World Production of Cobalt in 1970 (thousand tons)

Zaire	13 955	Zambia	2 177	USSR	1 542
Canada	2 370	Cuba	1 542	Finland	1 179
World	21 986				

Source: MY 1970, Vol. 1.

energy reactors (Nimonic). Heating coils in electric radiators also contain small amounts of cobalt in addition to many other alloy elements. Cobalt is also used in the production of permanent magnets.

Finally, cobalt is an ingredient in paints and medicine. The radio-active isotope can be used in radio-therapy. Almost all cobalt is obtained as a by-product of ores worked for other metals. Leading producers are Katanga[20] in Zaire, Sudbury[21] in Canada, Outukumpu in Finland and Bou Azzer in Morocco.

Columbium

Columbium (Cb), outside the United States niobium (Nb), in very small concentrations (0.05%) in normal carbon steels lower the grain size of steel, thereby increase its strength. This material is used for machine parts, e. g. cogwheels. Small amounts (0.7%) in low carbon steels increase creep resistance, i. e. resistance to deformation at constant load and increased temperature. Sometimes the input for this purpose is 1% which means high costs and exclusive uses such as jet engines, gas turbines, rockets, and guided missiles. Chief producing areas are Africa (Zaire, Mozambique, Nigeria, South Africa, and Uganda). Some production also takes place in Bolivia, Brazil, Guyana and Malaysia. Lovozero in the Kola Peninsula is reported to have one of the largest deposits in the world but Soviet production data are not known.

Tantalum

Tantalum (Ta) retains its strength at white heat and is used in alloys for jet engines and guided missiles. It is resistant to corrosion by most acids and is used for chemical, dental and surgical instruments. It is mined chiefly in Zaire, Nigeria, Brazil and Malaysia.

Rare Earth Metals

The rare-earth metals[22] (scandium (Sc), yttrium (Y) and the fifteen elements in the lanthanide-series) have acquired economic importance, i. e. as alloy metals. They are sold as rare-earth metal (REM) or misch metal, which contains some 50% cerium (Ce). Separation of individual metals is very complex because of their great chemical similarity. Prices for individual metals are therefore high. In 1970 lutetium (Lu) was most expensive, US $ 6500 per pound or 15 times the gold price. Cerium was cheapest, US $ 30 per pound. Rare earth oxides (REO) are cheaper than the metals.

Expressed in REO equivalent the primary uses for rare-earth products were as catalysts in petroleum cracking (45%), for glass polishing (25%), and for ductile steel (15%). In metallurgy misch metal is used because of its high affinity for sulphur and the increased hardness of the REM-sulfides. The increased ductility make such steels attractive in shipbuilding and for pipelines.

The rare-earth metals have turned out to be less rare than originally thought. The Soviet Union has large deposits in her apatite ores in the Kola Peninsula which are mined for phosphate and 1% REO. REO from here are exported to the United States and other countries by Techsnabexport in Moscow. REO in 1970 were removed from the American list of strategic materials. United States has a large production at Mountain Pass in California. India has a huge deposit in the monazite sands, $(Ce, La)PO_4$, of Kerala, which are also the principal thorium ore. Monazite is obtained along with titanium and zirconium from beach sands in Australia. Many other deposits occur in several other countries (Brasilia, Malaysia), but most deposits have not yet been exploited.

Base Metals

Iron is the most important metal in our civilization, used in much larger amounts than other metals. However, the modern machine culture depends not only on iron and steel for a proper work but also on several base metals such as copper, zinc, lead, and tin. Light metals and some other elements used in small quantities are also very important.

Tab. 36 – The Most Important Metals in 1973

Metal	World production million tons	Price, dollar/ton	World production billion dollars
Iron	460	84	39
Copper	7.5	1780	13.3
Aluminium	12.5	580	7.2
Zinc	5.56	850	4.8
Nickel	0.64	3310	2.1
Lead	3.55	430	1.5
Tin	0.23	4850	1.1

Sources: World Bureau of Metal Statistics, Metallgesellschaft and Metal Week

Copper

Copper (Cu, cuprum, Latin form for Cyprium, i. e. metal from Cyprus) has been used for millenia. Tools, weapons and ornaments of copper and bronze (an alloy of copper and tin) were used before iron was known, the Bronze Age. But copper really came into its own with the Age of Electricity. Copper is a good conductor, can be drawn into wire easily, and resists corrosion. More than half of all copper is used as wire in electrical equipment. Most of the rest is used in copper alloys, chiefly brass (copper and zink), but also bronze and German silver (copper, zink and nickel). Some copper goes to roofing, plumbing, hardware, jewelry, and decorative objects.

Copper deposits are quite widely spread but six countries account for $^4/_5$ of the world's mine output. Concentration, smelting, and refining are also geographically concentrated. All four steps are dominated by a few multinational groups. The flotation concentrator is at the mine, a smelter may serve several not-too-distant concentrators while refineries are few in number and often are near the market. They usually receive blister copper, which is almost pure, over long distances. However, the mining regions increasingly have got their own refineries. Thus Chile's copper exports in 1970 to 69% was in the form of refined copper.

Flotation enrichment, bulldozers and excavating machines were prerequisites for profitable exploitation of low grade copper sulfide ores in the beginning of the century (Bingham, Utah; Globe, Arizona; Chuquicamata, Chile). Copper sulfides account for some 90% of the mine output of copper. The most important copper minerals are chalcopyrite ($CuFeS_2$) and chalcocite or copper glance (Cu_2S). Copper is reusable on a large scale. Some 60% of all copper is recycled, but with a long lag: the average life of copper in a given use is 40 years. Secondary copper accounts for 40% of world consumption.

Tests have been made in the United States to use underground nuclear blasts for exploiting copper ores with even lower metal content than the national average rate of 0.6% but the method may be more useful in oil and natural gas fields in which the Soviet Union is also reported to have had experience.

Copper refineries in contrast to steel and aluminium works produce a few standardized products which are traded in a world market. Likewise in contrast to steel and aluminum copper prices fluctuate strongly. The base metals are traded in the London Metal Exchange (LME) where prices are quoted daily, loco, for immediate delivery to and from the LME warehouse and futures, for delivery within three months. The steel corporations, which are national in scope, and the multinational aluminum corporation change their list prices relatively seldom and changes are moderate (administered prices). Since both buyers and sellers of copper are dominated by a few large corporations covered by long term contracts the extreme fluctuations in the 'world market price' refer to parts of world production only.

The domestic American producer price in 1969 on the average was 72% of the London spot price and in 1970 it was 93%. The lowest average monthly price in London was 74% of the highest in 1969 and in America 93%. The corresponding figures for 1970 were 59% and 88%.

The leading exporting countries (Chile, Peru, Zambia, and Zaire) in 1970 formed the Council of Copper Exporting Countries (Cipec).

North America

United States is the leading producer and consumer of copper. The Keweenaw Peninsula of Upper Michigan has produced copper since 1845, and until 1885 was the most important district in the country. The pure or native copper mined in the deep mines of this district have been almost exhausted. The place of the mines has been taken by the low-grade sulfide deposit at White Pine, which was opened in the late 1950s.

In the last decades of the 19th century large deposits of low-grade copper ore were discovered in some western states and these now produce more than 90% of the American copper. Butte, Montana, is reputed to have mined more copper than any other district in the world. Gold and silver were mined here from 1862 but from 1892 these metals, along with lead, zinc and manganese, have been by-products of copper. The large hill at Butte, a granite batholit, is honey-combed with the mine tunnels of the Anaconda Corporation, built to exploit the rich vein deposits in the granite. However, most of the copper now originates in the low-grade porphyry copper deposits worked in open pits. The porphyry coppers are typical of the remarkable metallogenic province in the Americas, from central Chile to Alaska, but are found also in other parts of the world. (Granite Street and Porphyry Street are main streets in Butte.) The smelter at nearby Anaconda long claimed to have the tallest smoke-stack in the world, but, in spite of this, nature within miles turned into a moonscape as a result of the sulphur dioxide emissions.

The largest copper-producing mine in the United States is the open cut at Bingham, Utah, mined by the Kennecott Corporation. Although the ore yields less than 1% copper the mine has been profitable. Many miners commute from Salt Lake City, some 50 km away. Bingham is also a major producer of gold, molybdenum and silver.

The leading copper-producing state is Arizona (53% in 1970) with numerous mines, among them Morenci, Ray, San Manuel, Pima and Siderrita. The Phelps Dodge group has large interests in Arizona as well as Kennecott and Anaconda.

Copper production in the United States takes place within a few large groups, all with headquarters in New York and mines in the West. The refineries are strongly oriented towards the market, primarily near New York and Baltimore. Other refineries are located at Tacoma, Washington; Great Falls, Montana; Garfield, Utah; and El Paso, Texas. Large buyers are a few corporations like General Electric, Westinghouse, and Western Electric. The latter is the producing unit of the American Telephone & Telegraph (ATT) within the Bell System. It has 207 000 employees and belongs to the fifteen largest industrial groups, a grouping that also includes General Electric and Westinghouse. The mining corporations are much smaller; the three already mentioned and American Smelting and Refining belong to the 200 largest industrial groups in the United States. Because of the low metal content of their ores, averaging 0.6%, American mines have an output that fluctuates with the world market price. When prices are low, the least profitable mines close down.

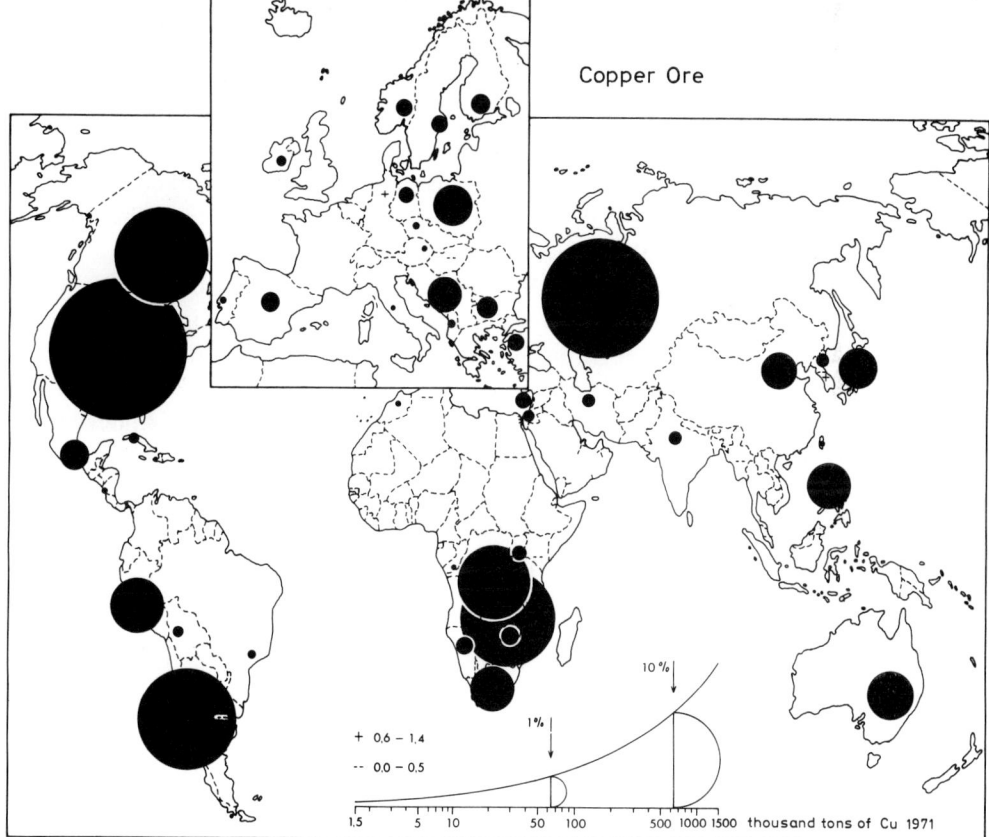

Fig. 28 – Copper mining in the world is dominated by three industrial countries: United States, Soviet Union and Canada. But also some LDCs are important producers: Zambia, Zaire, Chile, and Peru. Since the ICs dominate the engineering industry and therefore the market for copper and most other metals there is an important trade flow of metals from LDCs to ICs, tables 38 and 54.

In **Canada** the Sudbury district with its nickel-copper ores accounts for more than half the output. The Noranda-Rouyn district of Quebec near the Ontario border, where copper is associated with gold, ranks second and the Flin Flon area astride the Manitoba-Saskatchewan border ranks third. Copper is also mined at other places, e. g. in Newfoundland (Buchans Mine and Wabush) and Nova Scotia (Bathurst and Newcastle). As copper in Canada is a by-product of other metals, production is insensitive to fluctuations in world market prices.

South America

South America ranks high both in production and reserves. A copper belt stretches along the western slope of the Andes from northern Peru to middle Chile.

Chile was the leading producer until 1881 and now ranks fourth. It has the 2nd-largest known reserves in the world. Chilean ores are rather high-grade (2–4%) porphyry coppers, much higher than those mined in the US. The large copper mountain of Chuquicamata in the desert of northern Chile at an elevation of 3000 m, the world's largest copper deposit, was worked from 1916 by the Guggenheim brothers who brought the copper concentrate through the then recently opened Panama Canal (1914) to their refinery at Perth Amboy, New Jersey. Until the 1950s rich surface oxides were mined, but in 1952 large mills were completed by Anaconda to process the underlying sulfide ores. Everything must be brought in from outside for the mining operations and for the community of some 25 000 people.

The Teniente mine at 2 500 m altitude in a narrow valley of the Andes southeast of Santiago produces sulfide ore of 2% copper content under trying physical conditions. El Teniente is now the world's largest underground mine. El Salvador, high in the Andes 500 km south of Chuquicamata, was opened in 1959, the same year as the Potrerillos mine was closed down after being for 40 years the third ranking mine in Chile. The concentrates from El Salvador are

Tab. 37 – Refined Copper, Production 1973
 (thousand tons)

United States	2 085	Chile	415	United Kingdom	188
Soviet Union	1 300	Germany (W)	407	Poland	156
Japan	950	Belgium	372	Australia	150
Zambia	639	Zaire	322	Yugoslavia	138
Canada	498	Spain	189	China	118
World	8 490				

Source: SY 1974.

sent 30 km to the smelter at Potrerillos. New mines were opened in the early 1970s at Rio Blanco northeast of Santiago (the American Cerro group and CODELCO) and at Sierra Gorda near Antofagasta. During the Frei administration the government in 1970 took over 51% of the shares in foreign-owned copper mining subsidiaries through the state owned copper company (CODELCO). In November 1970 the Marxist leader Allende became President and soon the large copper mines, largely owned by American corporations, had been nationalized. In November 1973 the Allende government was unseated

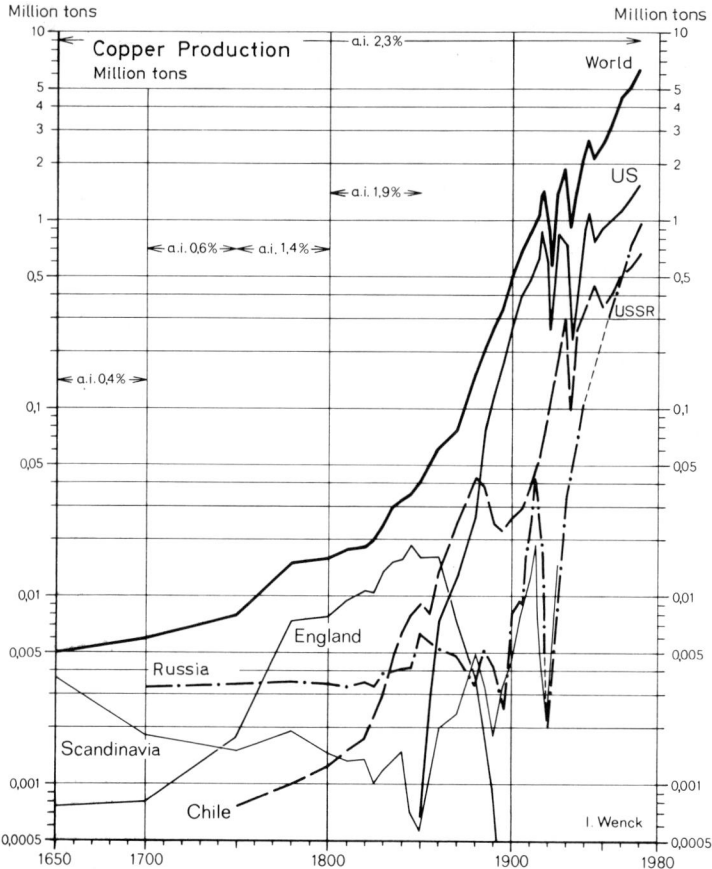

Fig. 29 – Estimates of the global metal production in the distant past are subject to wide margins of error. The annual rate of increase before the industrial revolution was very small. World production of copper in 1650 (some 5000 tons) was of the same magnitude as in small present-day producers, say, Albania or South Korea.

The dominating copper mine in Scandinavia was Falun, but also Röros and Lökken south of Trondheim were in operation. In England, the leading mines were in Devon and Cornwall. These mines were producing already in pre-Roman times. Tin was obtained in the same area.

through a military coup. The copper mines remained nationalized, but compensation to the former owners was pledged.

Peru has produced copper for four hundred years and for decades Cerro de Pasco in the Central Cordillera at 4 300 m flourished as a mining community in a feudal agricultural society in which the Indians provided the labor force. But this and other mines in the Andes are now overshadowed by the Toquepala open pit at 3 100 m in the desert of southern Peru which started production late in 1959, with a sharp increase of Peruvian output as a result. Another large copper mine in southern Peru with a capacity of 150 000 ton copper a year is expected to start 1976. The Hague-based Billiton N. V. will operate the mine for a consortium which also includes four American groups.

Africa

Copper is produced at several places in Africa but primarily in the world's leading copper belt that stretches 450 km from northern Zambia northwest across the Katanga province of Zaire. The mines of Katanga are mostly open pits while those of Zambia are deep shaft mines. Ores are high-grade, 3–5%, and reserves are very large but the long distance to the sea coasts and the lack of local fossil fuel have been serious handicaps. The district has been served by the Beira-Benguela railroad, originally built to serve the copper belt.

Zaire – As an African mining, manufacturing and urban complex the copper belt in Shaba (Katanga) is surpassed only by the Witwatersrand of South Africa. Developer was Union Minière du Haut-Katanga, a mining company, which at the end of the 1950s employed 21 000 Africans and almost 2 000 Europeans.

Union Minière was established in 1906 with the Belgian holding company Société Général, founded in 1822 as a major equity owner. It had its headquarters in Elisabethville (Lubumbashi). The company is now operating also in other countries, e. g. in the Canadian copper-nickel-field at Thierry, Ontario. It also has a subsidiary in Brazil, a country with a great potential for future operations.

The Congo became independent in 1960 and seven years later the copper mines were nationalized. The state-owned company is known as Gécomines (Générale Congolaise des Mines). Sales have been handled from Belgium by a subsidiary of Société Général, SGM (Société Général des Minerais), but from 1977 Gécomines plans to have its own sales organization at Kinshasa.

Most of the copper (220 000 t) moves the national route by train to Ilebo (former Port Franqui), by boat to Kinshasa, and by train to Matadi near the mouth of the Zaire River. In the Belgian period tariffs and rates were manipulated to induce the traffic to stay within the Congo. The Beira-Benguela railroad used to take 120000 t to Lobito and 50000 t to Beira. Some 80000 tons move by truck to Dar es Salaam and Mombasa. The railroad through Rhodesia is important because it supplies the copper region in Shaba (former Katanga) with 200000 tons of coal from Wankie.

Zambia – The Copperbelt of Northern Zambia, covering some $110 \cdot 55$ km, accounts for 95% of the export revenue. The copper content is high, on the average 3.4%. The mines employ 48000 Zambians and 5100 Europeans (expatriates). The largest mine, Nchanga, is the second largest copper mine in the world.

The Zambian mines have long been operated by subsidiaries of the South African group Anglo-American Corporation and the US-based American Metal Climax (Amax). However, since 1969 the state has had controlling interests in Nchanga Consolidated Copper Mines (NCCM) and Roan Consolidated Mines (RCM). In 1974 NCCM became independent with a director appointed by the Zambian government; a state-owned sales organization, Metal Marketing Corporation, was established.

The Copperbelt is served by the narrow gauge Beira-Benguela Railroad, which originally opened up the Copperbelt, but the changed political geography of Africa in 1970–75 led to the construction of a new railroad from Dar es Salaam, built by the Chinese Government. However, traffic congestion in the Tanzanian port may lead to only limited use of this outlet for the Zambian copper export. In addition, political events have changed the status of both Beira and Benguela (Lobito).

USSR

The Soviet Union ranks second among the copper producing countries. The Urals are no longer the dominating copper area, but several mines along the eastern flank of the mountains still make this one of the two leading copper regions. Most important is the Sverdlovsk area with the large mine at Degtyarsk. Complex copper pyrites dominate in the Urals. They also contain zinc, gold, silver and cobalt. Kazakhstan is the other leading copper district; the copper sandstone at Dzezkazgan is reputed to be the second largest copper deposit in the world. Another mine is operated at Kounradskiy north of Lake Balkash and a third area the lead-zinc mines of Leninogorsk near Ust-Kamenogorsk.

At Norilsk in northern Siberia copper is a by-product of a nickel ore from which platinum is also obtained. Many more recently discovered copper deposits in the Soviet Union are far from existing centers of population and manufacturing. Heavy investments in transportation are required whenever such deposits are developed. This is a common problem in the world's largest country with its low density of population.

144 Metals and Other Minerals

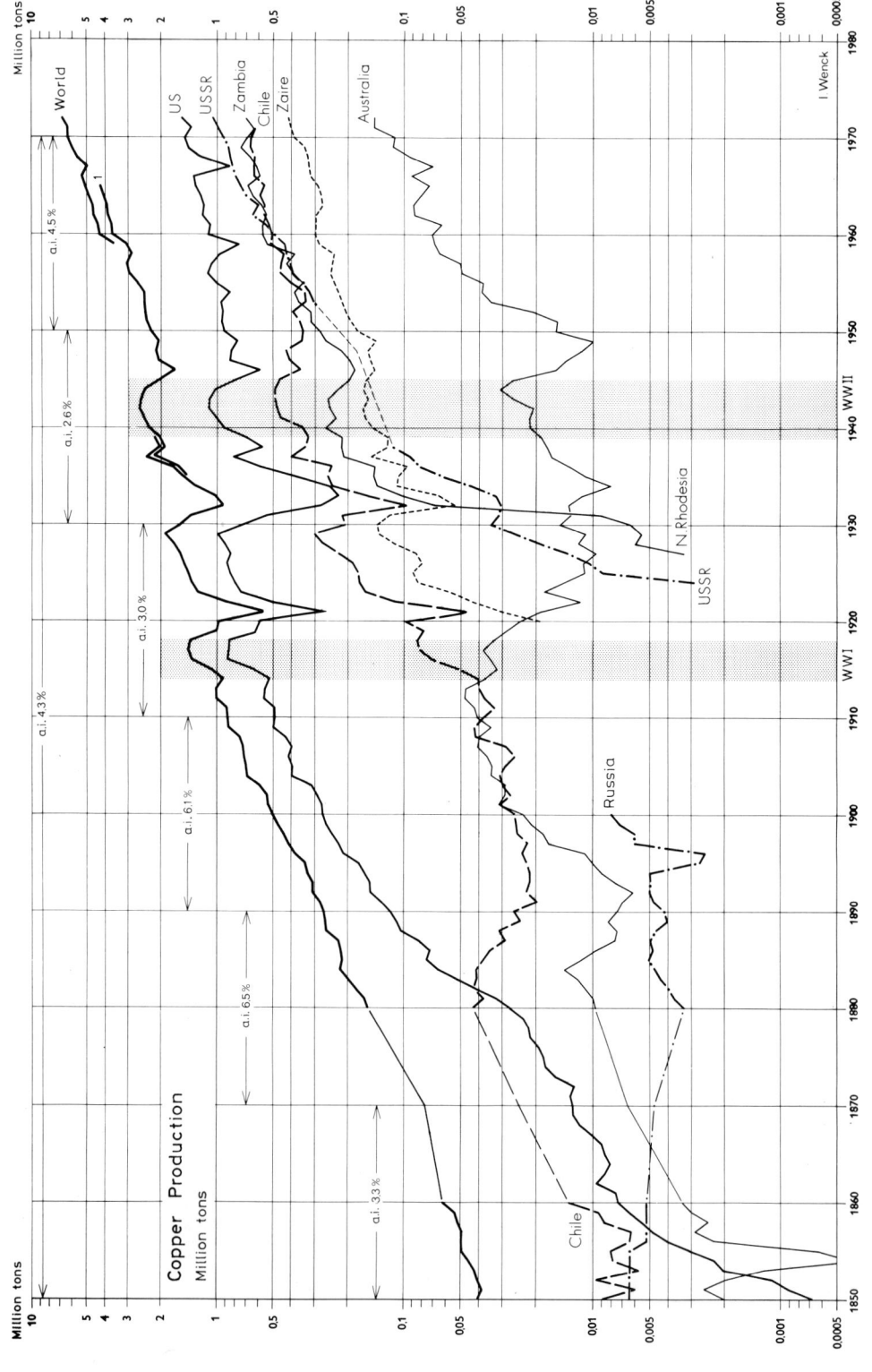

Mongolia – The large opencut at Erdenet 320 km northeast of the Mongolian capital Ulan Bator is planned to produce 16 million tons of copper and molybdenum ore a year from 1978. Mongolia is going to repay the Soviet investments with molybdenum and copper. The new town of Erdenet with 60 000 inhabitants will be the second largest in the country. A new railroad will connect the mine with the Soviet railroad net.

Australia has a large copper mine at Mount Isa, Queensland, and small production at several other mines. Mount Isa reserves average 3% copper, which makes it a very profitable mine. In the early 1970s a new large mine was opened in Bougainville in the Territory of Papua-New Guinea. PNG in 1975 became independent with guarantees for sizable Australian subsidies. The colonial administration was equity owner (20%) in the new mining company and its shares were taken over by the new government. The ore was found in 1964 after comprehensive prospecting by the Australian mining groups CRA and New Broken Hill.

Tab. 38 – International Trade in Copper 1970 (SITC 682) (million US dollars)

Imports		Exports	
Germany (W)	938	Zambia	950
Great Britain	681	Chile	949
France	542	Belgium	626
United States	532	Zaire	494
Belgium	530	Canada	454
Japan	484	United States	359
Italy	478	Germany (W)	335
Netherlands	184	Great Britain	304
Sweden	156	Peru	230
Total (excl CC)	5 610	Total (excl CC)	5 762

CC = Communist countries.
Source: YITS 1972–73.

Europe mines less than 5% of the world's copper. The Falun mine in Sweden, the mines at Mansfeld in Germany and the Rio Tinto mines in Huelva, Spain, famous in the history of copper have been closed or are small producers. Rio

Fig. 30 – Among copper countries with large production increases in recent years are the Soviet Union and Australia. The rather low growth rate in world production of copper, in a period of rapidly expanding electricity consumption, is, among other things, a result of substitution: aluminum lines can be used instead of copper lines.

Tinto, which in modern times have had strong British interests, has been mined for 3000 years, first for its gold and later for its copper and sulfur.

Leading producers in Europe are Yugoslavia with the Bor field, which before nationalization was developed by French interests, and Zajecar, which also has a refinery, Poland, with mines and smelters primarily at Legnica (Liegnitz) and Lubin near Wroclaw (Breslau), and Bulgaria, where Pirdop, Vraca, Pazardzik, and Panadjuriste account for a rapidly expanding production and where Pirdop has the smelter. The Nordic countries are also copper producers. In Finland the state-owned Outukumpu Company employs 10000 people and obtains copper, zinc, nickel, cobalt, sulphur and iron in ten mines of Carelia and northern Ostrobothnia. The chalcopyrite deposit of Outukumpu was discovered in 1910 and started production in 1913. The Imatra hydroelectric power station attracted the copper refinery but in the 1950s this activity was moved to Harjavalta and Pori on the Kokemäen River. The company's own flame melting method is used here as well as in many licensed copper works abroad. Most of the copper for the refineries is imported. The same holds for the zinc refinery at Kokkola. In Sweden the smelter and refinery is at Rönnskär serving the mines of the Skellefte field, which produce copper, zinc, lead, and gold. In Norway the main mine and smelter is at Sulitjelma and the refinery at Kristiansand.

Tab. 39 – International Trade in Ores of Non-Ferrous Metals 1970 (SITC 283) (million US dollars)

Imports		Exports	
Japan	1057	Canada	766
United States	522	Australia	272
United Kingdom	362	United States	239
Germany (W)	305	Philippines	202
France	195	Jamaica	185
Norway	180	Cuba	172
Canada	171	Bolivia	150
Belgium	131	New Caledonia	116
Netherlands	73	Indonesia	111
Total (excl CC)	3361	Total (excl CC)	3025

CC = Communist countries.
Source: YITS 1972–73.

Tin

Tin (Sn, stannum) has a low melting point, is malleable and ductile. It forms alloys easily, resists corrosion, is nontoxic, and makes excellent solder. Tin is used in most electric and electronic equipment but its principal use is as coating for steel, tin plate. Electrolytic tin plating allows extremely thin coating, some $1/200$ mm in thickness. Thus tin cans are more than 99% steel and only a fraction of one per cent tin. However, tin cannot be removed from steel during the remelting process, but instead is slowly enriched in the recycled scrap causing problems for steel shops depending on scrap as their raw material. Important tin alloys are: bronze (copper and tin), soft solder (tin and lead), babbitt metal (copper, zinc, arsenic, antimony, and tin) and type metal (lead, antimony, and tin). Compared to other metals world production of tin has grown very moderately, table 26.

The common tin-bearing mineral, cassiterite or tin oxide (SnO_2), is now primarily obtained from alluvial (placer) deposits. The primary (vein or lode) deposits are more expensive to work.

Southeast Asia mines $2/3$ of the World's tin in hundreds of placers and some underground mines. The tin-producing area includes Yunnan in south China, southern Thailand, the Malay Peninsula, and the Indonesian islands of Bangka, Billiton, and Singkep. Outside this area only Bolivia, the Soviet Union, and Zaire are of any importance.

Malaya with several hundred active mines is by far the largest producer; tin mining and smelting are basic industries in the economy of Malaysia. Large deposits are worked with huge dredges and many small deposits with gravel pumps and other less costly equipment. In Indonesia, Bangka produces almost $2/3$ and Billiton most of the rest. Some deposits are worked in submerged parts of river deltas with large sea dredges. Originally the concentrated cassiterite ore was shipped to Europe to smelters and refineries in Britain and the Netherlands but large smelters at Butterworth, Penang and Bangka now process most of the region's ore.

Tab. 40 — World Production of Tin 1970
(metal in produced tin concentrate, thousand tons)

Malaysia	73.8	Indonesia	19.1	Zaire	6.5
Bolivia	30.1	Australia	8.9	Brazil	4.3
Thailand	21.8	Nigeria	8.0	Argentina	2.3
Total	186.4	Data missing for the Soviet Union			

Source: SY 1972.

Bolivia — The tin mines of Bolivia are in a plateau and mountain area between 3500 and 6000 m above sea level, several hundred kilometers from the coast. Inhospitable climate, transportation difficulties, and lack of fuel are basic problems in addition to those posed by the complex association of minerals,

148 Metals and Other Minerals

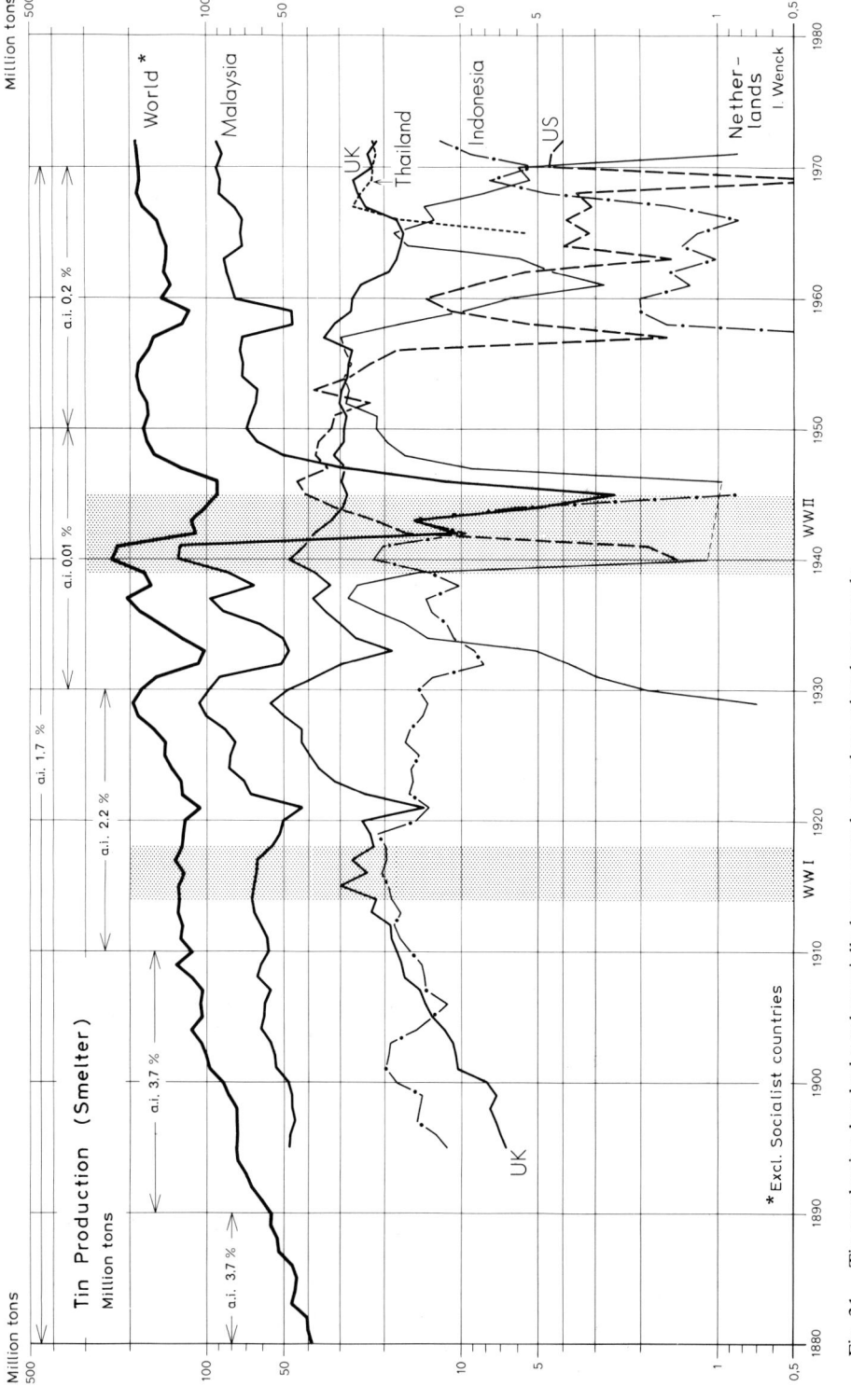

Fig. 31 – Tin production has had a substantially lower growth rate than other base metals.

which occur as narrow and irregular veins in solid rock. The only mines of importance are Catavi, Colquiri and Huanuni. Because of high costs Bolivia is a marginal producer of tin which cannot compete with Southeast Asia. But it is the only producer in the Western Hemisphere. In the Bolivian economy, tin is very important, by far the leading export item.

Tab. 41 – International Trade in Tin 1970 (SITC 687)
         (million US dollars)

Imports		Exports	
United States	190	West Malaysia	329
Japan	102	Thailand	79
Germany (W)	56	United Kingdom	58
France	38	Nigeria	46
Italy	28	Bolivia	35*
United Kingdom	24	Netherlands	22
Total (excl CC)	572	Total (excl CC)	640

* Estimated value for 1971.
  CC = Communist countries. China is an exporter of tin, the Soviet Union is a rather large producer.
Source: YITS 1972–73.

Africa is also an important tin mining area, primarily Zaire and Nigeria. In Zaire, tin is obtained in the eastern mining district as one of many important metals. The Nigerian tin is produced in the Jos Plateau in the northeast where the dry season poses problems for placer mining. Both Zaire and Nigeria gained prominence during World War II when the mines of Southeast Asia were taken over by the Japanese.

USSR – Until the 1930s, the Soviet Union had no production of tin, but later scattered deposits have been developed in the eastern border regions and in the North East. The Transbaikalia deposits are found southeast of Chita, near Komsomolsk and north of Vladivostok, those in the North East near Verchoyansk and Pevek. The Soviet Union now ranks with Africa or Bolivia as a tin producer.

Zinc

Zinc (Zn) is used for coating iron or steel (galvanization), which makes it more corrosion resistant than ordinary iron or steel, in alloys such as brass (copper and zinc) and in die-casting alloys which can also have an aluminium, copper, tin, or lead base, and as an element in voltaic cells. Zinc oxide serves as a paint

150   Metals and Other Minerals

pigment and enters into various chemical and medical preparations. Zinc used for galvanizing or as zinc oxide cannot be recovered. Secondary zinc accounts for only some 20% of the metal consumed.

Zinc is rarely mined alone but is usually associated with lead and often also with copper, gold, or silver. It is a rather common metal, found on all continents and in many countries. Production is no longer dominated by one or a few regions.

United States – The dominant position of the United States as a zinc producer has gradually been lost. It turned out 62% of the world's total in 1920 and only 9% in the early 1970s.

In the interwar years, three areas were of equal importance: 1) the western mountain states, 2) the three-state American corner of Missouri, Oklahoma, and Kansas, long the world's leading zinc area, and 3) states east of the Mississippi, primarily New Jersey (Franklin Furnace). In the early 1960s the western states accounted for 40% and the tri-state area for only 2%. The eastern states produced more than half but New Jersey discontinued production in 1958 when Pennsylvania resumed operations at the Friedensville mine which had been closed for 65 years. Tennessee has become the leading zinc-producing state, followed by New York, Colorado, Idaho and Missouri.

Canada – In Canada the Sullivan mine at Kimberley, British Columbia, is one of the world's largest lead-zinc mines. It also produces silver. The Mattagami area straddling the Ontario-Quebec border between James Bay and Georgian Bay is another large zinc producer with many zinc mines. The Kidd Creek-mine, 25 km north of Timmins, is a large zinc, copper, lead, silver, and cadmium mine. A third important mining area is south of Bathurst in northern New Brunswick. Zinc is one of many by-products at the copper mines of the Flin Flon region of Manitoba and Saskatchewan.

Tab. 42 – World Production of Zinc 1970
         (Zn content, thousand tons)

Canada	1 239	United States	485	Mexico	266
USSR	610	Peru	321	Poland	242
Australia	496	Japan	280	North Korea	130
Total	5 520				

Source: SY 1972.

Australia – The largest ore body in the world is Broken Hill, Australia, which has been mined by several large companies and has played a prominent role in

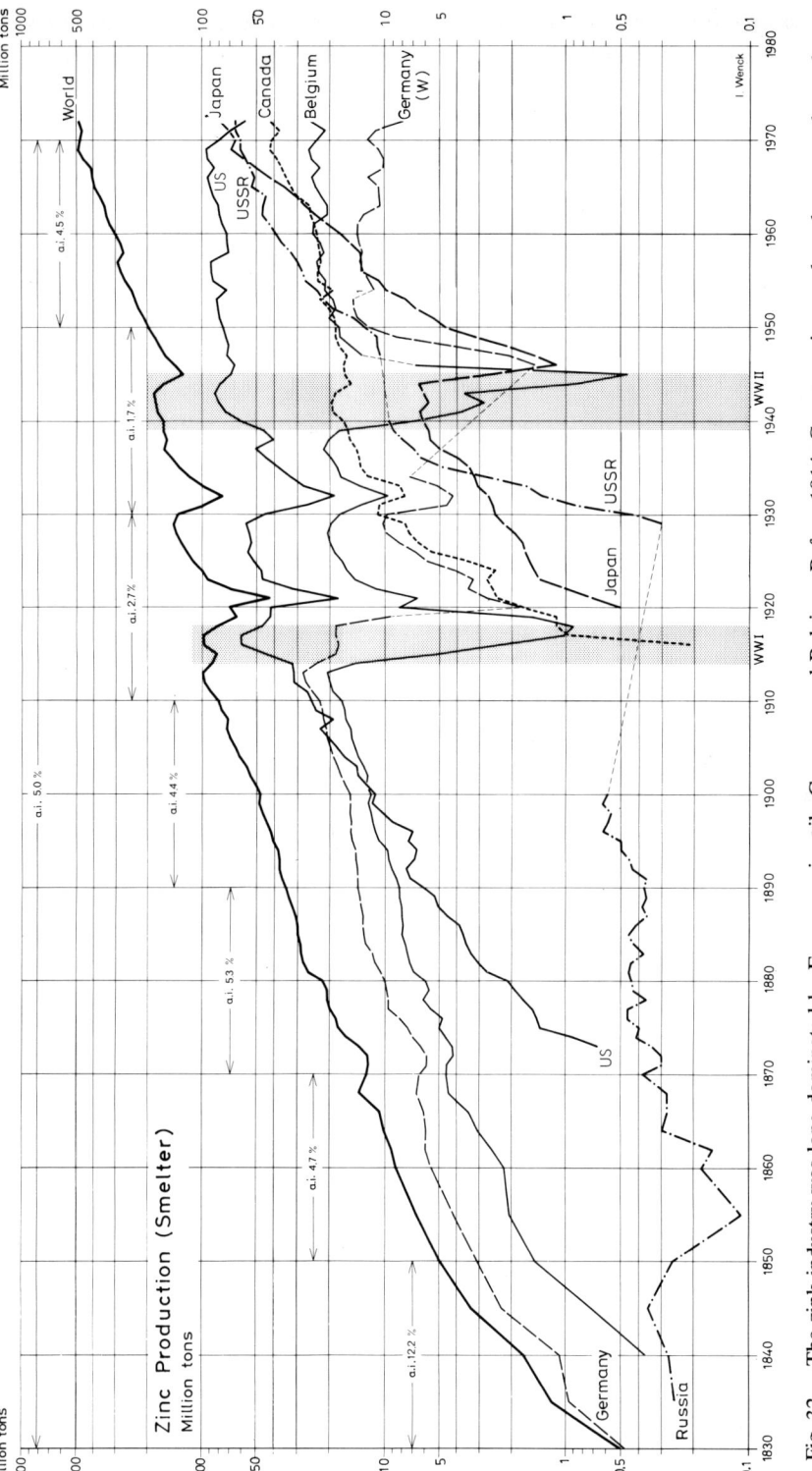

Fig. 32 – The zink industry was long dominated by Europe, primarily Germany and Belgium. Before 1914, German mines and smelters were located in Upper Silesia. After the war, some of these mines and all smelters were in Poland. A new smelter was built in the 1930s in Magdeburg to treat the ore from Upper Silesia. The Belgian smelters were based on imported or from, among other areas, Southern Europe and Sweden

the economy of the country since it was opened in 1885. The largest corporation in Australia, Broken Hill Proprietary Co. (BHP), derives its name from the mining area.

In the beginning silver was mined at Broken Hill. Due to the silver lobby in Washington and the Sherman Act silver fetched a high price in the world market. When in 1893 the United States and India stopped supporting the silver price, Broken Hill ran into its first crisis. In the beginning of the 1900s, when the easily processed oxide ore in the central part of the ore body had been almost finished, BHP decided to switch to the iron and steel industry, in which it later got a monopoly position.

Mining at Broken Hill now is in the hands of four companies, the old North Broken Hill and Broken Hill South as well as Zinc Corporation and New Broken Hill. The latter two, which mine the dominating southern part of the ore field, are parts of the London-based multinational group Rio Tinto-Zinc Corp (RTZ), whose Australian subsidiary is known as Conzinc Riotinto of Australia (CRA). The latter company has taken an active part in the mineral rush in Australia since the late 1950s. Broken Hill is known not only as a nursery of corporate managers and a place where important innovations were made in the flotation technique but it has also played an important part in the Australian trade union movement.

Other producing regions – The Soviet Union ranks among the leading zinc producers but production data by mine are not known. Zinc mines are found primarily in remote Asiatic parts of the country. Most important is Kazakhstan with Leninogorsk (zinc-lead-copper) and other mines in the Altay region and Chimkent in the Tien Shan Mountains. Tetyukhe northeast of Vladivostok and Ordzhonikidze in the Caucasus are other mining districts. Zinc refineries are located at places with low electricity costs to which the mineral may be hauled very long distances: Ordzhonikidze (hydro), Konstantinovka in Donbass and Belovo in Kuzbass (coal), Chelyabinsk (lignite) and Ust-Kamenogorsk (hydro).

In Latin America, Mexico and Peru are leading producers. Mexico has many mines, primarily in the states of Zacatecas and Chihuahua, but the Peruvian output is dominated by Cerro de Pasco.

The Kipushi mines of northern Zambia are the leading producers in Africa. Also Morocco and Namibia have zinc mines.

Japan has a significant production from several mines in Honshu and Hokkaido.

In Europa Poland and Italy are leading producers but also Spain, Greece, Germany (W), Sweden, and Finland have mines. The largest Nordic producer in 1974 was Greenland, politically part of Denmark and geographically belonging to North America. The new mine turned out zinc and lead.

Lead

Lead (Pb, plumbum) is one of the earliest known metals. The Chinese used lead coins some 2000 B.C. The silver-lead mine at Laurion near Athens was exploited about 1200 B.C. The water conduits of lead at Pompeji and the lead roofs of the Venezian prisons are well-known. Lead is heavy (d 11.3), soft, ductile, and malleable. It alloys easily and is corrosion resistant. It has a low melting and a high boiling point and is not penetrated by short-wave radiation. The electric industry takes almost half the lead for storage batteries and cable coverings. The chemical industry is another large lead consumer: tetra-ethyl-lead, paint pigments, ceramics, plastics, bullets and insecticides. Lead is used in the construction industry and in alloys such as solder, bearing metals, and type metal. A use of increasing importance is as radiation shields in production of

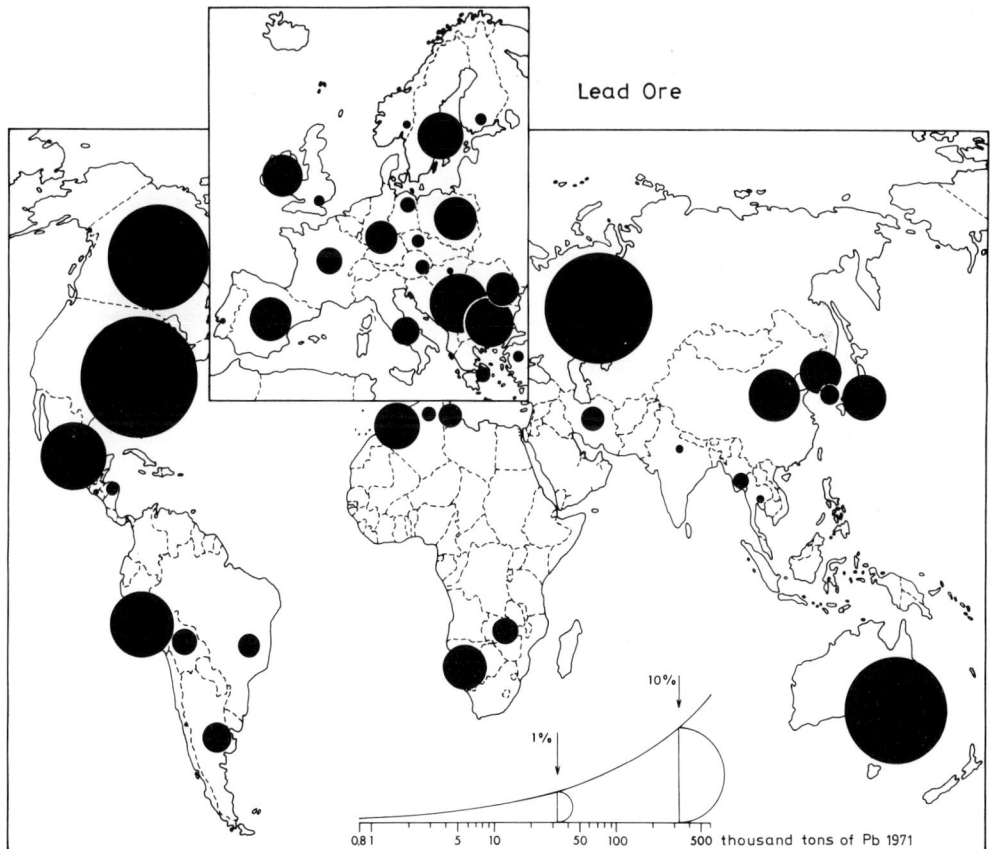

Fig. 33 — Lead ore mining occurs in many countries. Lead as a rule is associated with other metals, primarily zink, in sulfide ores.

nuclear energy. More than 40% of the lead used in the United States has been secondary metal; the recovery rate is high.

The most important lead minerals are the sulfide galena (PbS), which almost invariably contains silver and is the dominating lead ore, the carbonate cerussite ($PbCO_3$) and the sulfate anglesite ($PbSO_4$). Lead often occurs with zinc and as a rule lead ores also contain other metals.

United States was the leading producer 1892–1956 and is again in that position since 1969. In the meantime it was surpassed by Australia and the Soviet Union. Until the Second World War, America was a net exporter of lead, but more recently American mines have at times delivered only $1/4$ of the national consumption. The balance has been supplied by secondary lead and imports.

Mine output has long been dominated by the lead belt of southeastern Missouri, some 100 km south of St. Louis. Large deposits of low-grade (3%) galena ore are mined in large-scale operations. The Missouri mines, which contain pure lead ore not associated with other metals, have been worked since the French time (1725). In 1970 they accounted for $3/4$ of the American mine production.

In the Coeur d'Alene district of Idaho, lead is associated with silver and zinc in ore veins occuring in a much folded and faulted area. Some mines are about 2000 m deep. Lead production expanded after introduction of the selective-flotation process.

Canada – The Sullivan mine at Kimberley, British Columbia, some 140 km north of the Idaho border, is Canada's largest producer. The huge ore body is similar to that of Cœur d'Alene. The Northwest Territories became the leading lead-producing area with the opening of the Anvil lead-zinc mine in Yukon. The mines south of Bathurst in northern New Brunswick are also important lead producers.

Mexico is regularly the largest producer in Latin America and has at times attained the second place in the world. Numerous small deposits account for much of the output. As a rule the lead ore is associated with silver and zinc.

Australia – The drastic increase in Australian output of lead has been from old mines, primarily those operated by four major companies at Broken Hill in New South Wales, which in 1970 accounted for 62% of the production. Also Mount Isa, Queensland (33%), and Rosebery, Tasmania (3%), produce lead. The ore averages 11–17% metal in the Broken Hill lode which has had no equal in the history of lead mining. Lead and zinc were important export items from Australia even before the minerals boom of the 1960s.

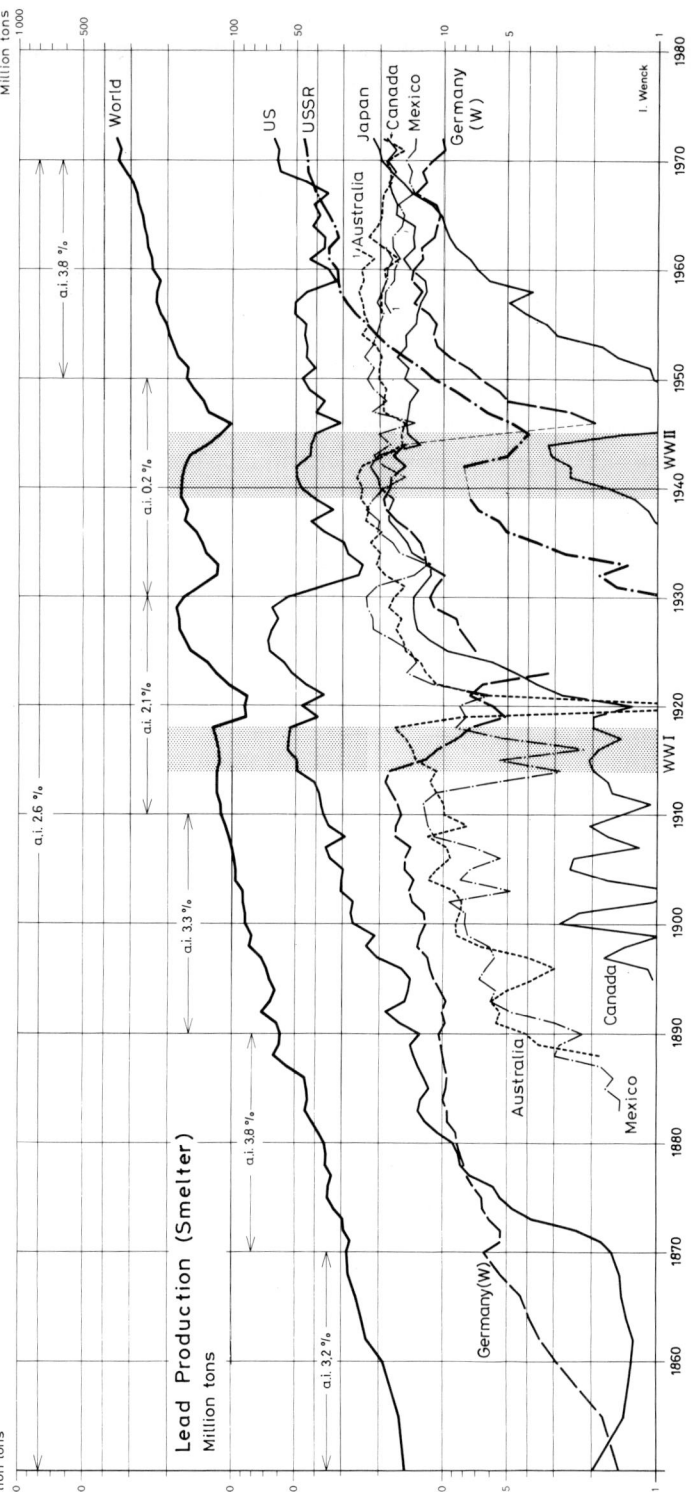

Fig. 34 — The production of primary lead has registered a moderate rate of increase in comparison with other leading metals. The smelters get their raw material in the form of concentrates from domestic and foreign mines and some per cent in the form of scrap. In the United States and some other industrial countries, smelters provide less than half the consumption of lead. The rest is obtained in the form of scrap.
Europe has traditionally had a larger smelter production than ore output of lead. The smelters have been market oriented. Within the market region, they have been located with access to cheap fuel (coal, brown coal).

156 Metals and Other Minerals

USSR – The large lead production of the Soviet Union comes almost exclusively from the Asiatic part of the country and is usually associated with zinc mining.

Europe – In Europe, Yugoslavia and Bulgaria are leading producers, but also Sweden, Spain, Poland, and Ireland have a notable production.

Mercury

Mercury or quicksilver (Hg, hydrargyrum) is unique among the metals since it is liquid at ordinary temperatures. The melting point is $-39°C$. It is toxic, has high density and electric conductivity. It is used in electric apparatus, in thermometers, barometers, and other scientific instruments, in pesticides, pharmaceutical preparations, and dental fillings, in reflecting surfaces of mirrors, and as a laboratory catalyst. Although minor in tonnage used, mercury is of strategic importance. Almost all mercury is obtained from cinnabar (HgS), a sulfide ore found in volcanic areas. Many experts believe that mercury will be the first scarce metal to be exhausted. That day will be reached within a matter of decades.

Italy and **Spain** long dominated world production and still rank among the leading producers. Two mines at Monte Amiata near Siena in Tuscany account for most of the Italian output. The state-owned Almadén mine in the center of Spain has much richer ore than other major deposits and after more than two millenia of mining still contains the world's largest reserves.

USSR – The leading Soviet mines are at Khavdarken in the Kirgiz Republic and at Chagan-Uzum in central Siberia near the border of Mongolia.

North America – Other large mercury producers are Mexico with mines southwest of Mexico City near Huitzuco and United States with mines in a zone from southern Oregon reaching 650 km into California and Nevada. American ores are low grade and mining costs high. Output fluctuates greatly with changes in mercury prices. In 1943 wartime demand made 200 mines produce 52 000 flasks of 76 pounds each while in 1950 only 16 mines were operating, producing a postwar low of 4 500 flasks. The American 1943 production equalled the postwar Italian or Spanish output.

Pollution of the environment. In the 1960s it became obvious that mercury like certain other heavy metals (lead, cadmium, zinc, chromium, copper, barium, and selenium), which occur in nature in small concentrations, through the influence of modern technology in some contexts may occur in concentrations unacceptable to humans and animals.

Mercury-disinfecting of seed to prevent fungus, which in its turn hinders the seed from sprouting, was introduced in Germany about 1914 and became common after 1920. By eating fowl that have eaten such seed people have been mercury poisoned. Others have been poisoned by eating eggs. The worst mercury catastrophe occurred in Iraq in 1972 when 6500 people were brought to hospital after eating home-baked bread which by mistake had been baked from mercury-disinfected seed from Mexico. The number of deaths was listed at 459. This type of accidents are more likely to happen in underdeveloped countries, where the control system has not been fully enforced.

In the industrial countries mercury is used in many manufacturing processes (pulp and paper, plastics and so on) and large quantities of mercury end up in lakes and rivers via industrial effluents. By eating fish from such waters people may get unacceptable concentrations of mercury in their bodies. The Minamata catastrophe in Kyushu, Japan, with 50 deaths and hundreds of cripples was the alarm clock. The cats of Minamata suddenly started to jump and dance and finally dived into the sea and drowned. Even the fishermen and their families showed similar symptoms. The illness was first described as *fūdo-byō*, i. e. local sickness, which formerly was a common name for unknown diseases. An investigation showed that mercury from a local fertilizer plant by way of the fish had ended up in cats and human beings.

Around 1970 some industrial countries passed laws which regulated the use of mercury. Fishing was forbidden in some waters. These measures had an impact on the world consumption of mercury.

Tab. 43 — World Production of Mercury 1970

USSR	1 655	Mexico	1 043	Yugoslavia	533
Spain	1 570	United States	941	Turkey	324
Italy	1 530	China	690	Japan	202
Total	8 871				

Source: SY 1971.

Precious Metals

Precious metals, including gold, silver and the platinum group, have important industrial uses. In addition, gold and silver have long been monetary standards, gold more commonly in the Occident and silver in the Orient, especially China.

Both gold and silver are prehistoric metals. For geological reasons gold was originally more common than silver. Native gold could be obtained with primitive methods (panning, mechanical working of gold veins) while silver required more advanced mining and smelting methods. The largest silver mine of Antiquity was Laurion south of Athens which employed 20 000 slaves and played a central role in the economy of Athens.

A special measurement system is used in the Anglo-Saxon world for the precious metals, the troy system, named for the French town of Troyes. A troy ounce is 31.103 g in the cgs-system.

Gold

Gold (Au, aurum) has two distinct uses, as an international reserve asset and as commercial metal. The reserve is kept by central banks in ingots of the metal (bullion). Some countries still make gold coins. Some $3/4$ of the commercial gold is used in jewelry, the rest in dentistry, as a conductor in the electronics industry and so on. Pure gold is too soft for jewelry. As a rule it is alloyed with copper and other metals. The fineness of gold is measured in karat, 24 karat being pure gold. Gold of 14 karat contains 58% gold and nobler metals (platinum), 25–32% copper, and some silver, zinc, or nickel. Most gold is obtained as native gold, which is not pure but almost invariably contains some silver. In the past, gold was primarily produced by placer mining but for many decades the chief source has been quartz veins or sulfide ores, often obtained in shaft mines. Much gold is obtained as a byproduct or coproduct in mines operated for other metals.

For the early years of the 16th century world production has been estimated at 200 000 troy ounces. The search for gold was a major driving force during the era of great discoveries. Until the middle of the 19th century production increased steadily but slowly. More than 80% of the gold was produced in South and Central America. The California gold rush in 1849 and the rush to Victoria, Australia, in the 1850s brought about a great rise in output which was 40 million fine ounces just before the First World War. A great drop during the war (to 26 million) was followed by recovery. In the mid-1960s production was about 45 million ounces and in 1970 a record of 47 million ounces was reached. Gold scrap is an increasingly important source of commercial supply, almost 40% in the United States.

The gold price was unchanged, US $35 an ounce, between January 1934, when Roosevelt pegged gold at 59% of its former gold value and August 1971 when Nixon discontinued the American pledge to buy dollars from foreign central banks at the price of $35 an ounce. An unchanged gold price for almost 40 years of inflation had made gold mining an unprofitable activity. After

August 1971 followed a time of floating exchange rates for most industrial countries and two American devaluations ($38 an ounce in 1972 and $42.22 an ounce in 1973).

The two separate uses of gold in 1968 became two separate markets. Sale of gold from the central banks to the free market was no longer allowed. The fluctuating gold price in the free market, set at auctions primarily in London and Zürich, in December 1974 reached the record level $197 an ounce. The official IMF-price at the same time, after the two American devaluations, was US $42.22, which in reality meant that gold reserves had been frozen and that gold no longer could be used in international transactions.

According to a law of 1933 Americans were not allowed to own gold in bullion. This law was changed on December 31, 1974. The American government, who wants to demonetize gold, made a declaration that it would auction parts of its gold reserves, which were some 274 million ounces worth US $11.6 billion according to the official price, $42.22 an ounce.

From December 31, 1974, Americans were also allowed to own gold coins. The South African Krugerrand, which weighs one ounce, sold at more than 200 dollars in November 1974 at the exchange in London. In ten months 1974 South Africa sold 3.9 million coins; 18% of its gold production was coined.

South Africa in 1970 accounted for $^3/_4$ of the known world output, almost all from 46 deep shaft mines. South Africa's share of world production has grown steadily from less than half in the 1950s. Many mines produce uranium as a byproduct. The world's largest gold reef, the Witwatersrand or the Rand in Transvaal, where gold was found in 1886, has attracted the largest concentration of population in South Africa. In addition to Johannesburg, the economic capital of the country, a string of mining and industrial towns (Springs, Benoni, Krügersdorp and others) have grown up in the mining area. Operations started in the surface outcrops. The gold-bearing strata dip steeply to the south to depth of at least 3600 m. The deepest mine, 2700 m below the surface, has serious temperature and rock pressure problems. Since 1949 production in the Welkom-Odendaalsrus-area in the Orange Free State has increased its share.

Gold mining is still very important in the South African economy in spite of rapid expansion in other sectors of the mining industry. But profits were low in many mines when gold could be sold only at the monetary price of US $35 per ounce. From 1968 marginal mines were eligible for government subsidies. Such help to gold mines occurred also in other countries, e. g. in Canada. Gold prices remained stable from 1933 to 1971 while labor costs increased and lower ore grades had to be worked in deeper mines.

The drastic increase in gold prices on the free market has strikingly improved the profitability of gold mining. Gold shares went up 500% from the beginning of 1972 to early 1974 in Hollard Street, Johannesburg. Almost 40% of the

160   Metals and Other Minerals

South African gold production is accounted for by the Anglo-American Corp, the 'Oppenheimer Empire', which has producing subsidiaries in several countries of southern Africa.

USSR – The most important gold mining area of the Soviet Union is the Yakutsk region of northeastern Siberia in the valleys of the Aldan, Indigirka and Kolyma rivers, where huge dredges work rich placers. Gold is mined at several other places in Siberia. The major lodes are mined east of Lake Bajkal.

In the Soviet Union data on the production of precious metals and some other nonferrous metals are considered state secrets according to a law of 1956. This runs counter to the interests of the international banks that need the data to judge the credit standing of the country. The Bank for International Settlements in Basel (BIS) reports that the Soviet gold sales in 1973 were 330 tons (settlement of the large grain imports) and in 1974 some 150 tons. The American Central Intelligence Agency (CIA) in 1975 estimated the Soviet gold reserves at 2000 tons, much larger than the South African reserves, and the Metal Bulletin of London gave the Soviet gold production for 1975 at 440 tons or half the South African production.

The Soviet Union, South Africa and holders of large gold reserves like France and Italy plead for a greater role for gold in the monetary system and a return to the pre-August-15-1971-convertability of gold. As mentioned, the United States is on the opposite side.

Tab. 44 – World Production of Gold 1970*
         (million troy ounces)

South Africa	32.16	United States	1.74	Philippines	0.60
USSR	6.50	Ghana	0.70	Rhodesia	0.50
Canada	2.34	Australia	0.62	Japan	0.26
Total	47.36				

* World production (excl USSR and China) declined from 1288 tons in 1970 to 1135 tons in 1973 according to SY 74. All major producers reported declines except Japan that exceeded Ghana, and Papua-New Guinea that exceeded the Philippines.
Source: MY 1970, vol. 1.

Canada – Gold mining was of great importance for the development of northern Canada and for the national economy in the first half of the century. From about 1920 to the 1950s gold was Canada's most valuable mineral. But from 1960, when Canada was the second largest gold producer in the world, output has continuously declined to a mere 5% of world production in 1970. Yellowknife in the Northwest Territories and Red Lake are the major mines outside the dominant areas of eastern Ontario and adjacent parts of Quebec, among them Timmins, Kirkland, and Larder Lake. In Canada gold mines as a rule were subsidized by the government in the late 1960s; only one lode mine, Red Lake in western Ontario, operated without subsidies.

United States — In the United States the Homestake mine in the Black Hills of South Dakota has been the largest in modern times. The Bingham mine of Utah is second in importance. Here gold is a byproduct of copper. Nevada is second among American states with mines at Carlin and Cortez.

Silver

Silver (Ag, argentum) is used in jewelry and fine housefurnishings, e. g. tableware and in photography and for coating of German silver. Use of silver in coins has declined in many countries but it is still used in commemorative coins and medals. Among modern uses are electric and electronic apparatus. For coins and jewelry silver is alloyed with copper and nickel to become harder.

In earlier times, silver was a standard of value along with gold. Until the British acts of 1798 and 1816, which made gold the standard currency, all countries practiced bimetallism. Many countries abandoned bimetallism for the gold standard in 1873. In the United States the final step was taken in 1900. From the early 1870s the supply of silver increased rapidly with the tremendous production in Comstock Lode, Nevada, and other mines in the American West. More and more units of silver corresponded to one unit of gold; bimetallism became increasingly impractical.

For millenia the value of the two precious metals, gold and silver, was decided by commercial supply and demand and not by government decree. By peculiar coincidence the relation between the two metals long remained more or less constant: 15–16 units of silver were worth one unit of gold. However, the commercial relation always remained a variable while the legal one in times of bimetallism was a constant, often 16 : 1. Knowledgeable people could analyze the coins and hoarded those that were coined in the high-value metal and used those of the low-value metal. Bad money drove good money out of circulation (Gresham's law). Bad coins became the dominant currency while good coins were smelted and utilized for their metal.

In more recent times the relative value of currencies have been decided by governments who issue paper money which after legislation serve as means of payment (fiat money). As an international standard of value gold in the early 1970s was getting a competitor, 'paper gold' or special drawing rights (SDR). In the system of fixed exchange rates agreed upon in Bretton Woods 1944, which worked until 1971, US dollars and British pounds served as reserve currencies.

The Western Hemisphere has turned out some 80% of all silver ever produced and North, Central and South America still account for ³/₄ of the world's output. United States, Canada, and Mexico each produce almost 15% of the world output followed by Peru, USSR, and Australia. Over 50 countries produce silver but the six leading countries account for over ³/₄ of the total output.

About ²/₃ of the American silver is a byproduct of mines producing base metals and gold. Only in Idaho, the chief silver producing state with 42% of the American output, silver is the main element in the ore. Sunshine in the Cœur d'Alene district is the largest American silver mine. Arizona, Utah, and Montana all accounted for more than 10% in 1970.

In Canada silver is mined in the Timmins district and other metalliferous areas straddling the Ontario-Quebec border, in the Yukon area and at Port Radium in the Northwest Territories, at Kimberley and Flin Flon, and in several mines of the Maritime Provinces.

Mexico has many rich deposits in its western mountains scattered from north to south throughout the country. At times it has been the leading silver producing nation in the world. Mexico is the only country in which the bulk of its silver is obtained from ores in which silver is the most valuable element.

In Peru, Cerro de Pasco is still the principal source of silver with both shaft and open-pit mines. The company with the same name operates several other silver mines in central Peru. A predicted large expansion of copper production, primarily in southern Peru, during the 1970s will substantially increase the output of byproduct silver. Bolivia is a much smaller silver producer.

USSR – The Soviet silver deposits are located in Asia, primarily near the Chinese border, in Kazakhstan and western Siberia.

Tab. 45 – World Production of Silver 1973
(thousand tons of Ag in ores mined)

Canada	1.52	Mexico	1.21	Japan	0.36
USSR	1.28	United States	1.18	Germany (E)	0.22
Peru	1.25	Australia	0.61	Bolivia	0.16
Total	9.18				

Source: SY 1974.

Platinum

Platinum (Pt) and its group of metals (palladium, Pd, iridium, Ir, rhodium, Rh, ruthenium, Ru, and osmium, Os) are rare elements with important industrial uses, as a rule more costly than gold. Petroleum refiners, organic and inorganic chemical industry, manufacturers of electric and electronic equipment are the major consumers of the platinum metals. The Clean Air Act of 1970 is expected to boost American consumption of platinum if car manufacturers decide to meet the 1975 emission standards by using platinum catalysts. Over 90% of total production is accounted for by palladium and platinum.

Major producing countries are the Soviet Union with leading mines at Nizhniy Tagil in the Urals where platinum was discovered in the early years of the century, at Norilsk near the mouth of the Jenisej and at Pechenga in the Kola Peninsula. Platinum metals in 1973 accounted for $^1/_3$ of the American imports from the Soviet Union, primarily by Engelhard Minerals & Chemical Corp, the leading US producer of emission control equipment for cars.

South Africa is the other large producer with mines in the Rustenburg and Lydenburg areas of the Transvaal, Canada, a poor third among the world's producers of platinum metals, gets most of its output as a byproduct of nickel mining and refining in the Sudbury district.

Light metals

The line between light and heavy metals is drawn at density 4.0. Metals show a great spread in densities, between 0.534 for lithium and 22.5 for osmium. Lightness is a desired quality in construction materials, not only for airplanes and space craft but also for land and sea vehicles, containers and packages.

Titanium (d 4.5) is a border case but is included here since lightness is the main reason for its use in the aerospace industry. Since the strength of titanium is superior to that of aluminium thinner constructions are being used. A given titanium part is lighter than the corresponding part in aluminium.

Aluminium

Aluminium (Al) occurs abundantly in all ordinary rocks, except limestone and sandstone. Only oxygen and silicon among the elements are more common than aluminium (8.1%) in the Earth's crust.

History – Aluminium was first isolated in 1825 by the Danish chemist H. C. Ørsted. The method was improved by the French H. S.-C.Deville, who showed the metal, worth more than gold, at the Paris Exhibition 1854. Originally, cryolite (Na_3AlF_6) was used as a source. The only known deposit of cryolite is found at Ivigtut in Greenland. Mining was discontinued in the early

1960s when the mine had been emptied. This strategic material can now be made synthetically, but still in the early 1970s natural cryolite was sold from stocks.

The electrolytic method, by which aluminium is now produced from alumina, was discovered in 1886 by C. M. Hall (United States) and Paul Héroult (France), working independently. By mixing in cryolite the melting point of alumina is lowered from 2035 °C to 950 °C.

Hall made his discovery as a student (22) at Oberlin College, Ohio, and managed to interest a group of people in Pittsburgh (Captain A. E. Hunt and others) who in 1888 formed the Pittsburgh Reduction Company which later changed name and became the Aluminum Company of America. Héroult was of the same age as Hall. He went over the border to Neuhausen in Switzerland, where Aluminium Industrie Aktiengesellschaft was formed in the same year. The French name form is Alusuisse.

Pure alumina is made from bauxite, which is the dominating raw material for aluminum. The method was launched in 1889 by the German K. J. Bayer. It presupposes the Solvay method of producing cheap caustic soda.

Bauxite, named for the little French village Les Baux in Provence where the ore was found in 1821, contains 45–65% aluminum oxide or alumina (Al_2O_3). The claylike ore primarily contains aluminum hydroxide but also impurities in the form of silica, clay and iron oxides. It was formed by leaching under a humid tropical climate. Deposits now found in temperate regions, primarily Europe, were presumably formed when those areas had a warmer climate. Because aluminum components have low solubility they become concentrated under high temperature and precipitation when silica is leached out. Iron oxide as well as aluminum oxide end up in the concentrated residual. The lower the content of iron oxide, the higher the quality of the bauxite deposit. Bauxite is beneficiated at the mine where it is crushed, washed and dried.

Alumina – In the Bayer process alumina is obtained through mixing bauxite in powder form with a hot solution of caustic soda. This is pumped into pressure tanks where the aluminum hydroxide is dissolved from the bauxite.

The impurities of the ore are not affected by the caustic soda but are collected as 'red mud' in filters when the solution is pumped into precipitating towers, as high as six-story buildings. The aluminum hydroxide is here precipitated as fine crystals and the caustic soda is pumped back to the beginning station of the process, which thus is continuous.

The aluminum hydroxide is then heated in large rotary kilns. The white-hot hydroxide gives off water, and changes into aluminum oxide which does not reabsorb moisture from the air upon cooling.

During World War II the Bayer process was supplemented by Alcoa's Combination process, which permits the use of low-grade ores. The 'red mud' is here calcified with limestone and soda ash. Installations were built at East St. Louis and Hurricane Creek.

Aluminum − Hall and Héroult found that aluminum oxide is decomposed by the passage of an electric current if the oxide is dissolved in molten cryolite. The solvent is not affected. Furnaces are steel cells lined with carbon, containing molten cryolite. An electric current passes from hanging carbon anodes to the bottom of the cell (the cathode). The oxygen moves to the anode and escapes as carbon dioxide through the crust of the bath. The aluminum ends up at the cathode and is tapped from the bottom of the cell into a ladle and cast into pigs, each weighing 23 kg. An aluminum reduction plant or smelter has long batteries of cells, each producing some 340 kg a day. The pigs contain some impurities and must be remelted to obtain pure aluminium; alloying takes place during this process.

Economics − A ton of aluminum metal requires 4−6 tons of bauxite, which yield 2 tons of alumina. Before the bauxite leaves the mine it is beneficiated to remove moisture and impurities. As a rule more than 4−6 tons of raw bauxite must be mined for each ton of metal.

The Hall-Héroult-process, which in 1950 required some 22 000 kWh electricity per ton aluminum, 600 kg carbon electrodes, and 100 kg cryolite, in 1970 needed some 16 000 kWh. The electricity cost made up some $1/3 - 1/4$ of the total cost for aluminum.

Uses − Aluminum is the only light metal (d 2.7) produced in large quantities. In alloyed form it is strong (duraluminum) and is used in the construction of airplanes. The metal is resistant to air corrosion, it is a good conductor of heat and electricity, it reflects light and heat, it is non-sparking, non-magnetic and nontoxic.

Aluminum is used in the construction industry, in the manufacture of aerospace, sea and land vehicles, in household appliances, cooking utensils, machinery and electrical appliances, containers, irrigation pipe, television antennae, awnings and venetian blinds, and foil. An aluminum cable has 50% more strength than a copper cable and at the same time is 20% lighter. Fewer poles per mile are used (11 for aluminum and 15 for copper).

In the United States 22% of the 1970 production of aluminum went to the construction industry, 15% to the manufacture of transport vehicles, 15% to containers and packages, 13% to the electric industry, 9% to durable consumer goods, 6% to machinery, 12% to exports and 8% to miscellaneous uses.

Corporations – The three steps in the manufacture of aluminum have widely different distribution patterns. Production is financially integrated but geographically separated and controlled by six multinational corporations, three of them US based: Aluminum Company of America (Alcoa), Reynolds and Kaiser.

Alcoa was the most pronounced monopoly in the United States before World War II with 100% of the national production of aluminum oxide and metal. The Federal Government had long tried to break the monopoly and in 1940 helped Reynolds build a smelter at Longview, Washington, and a combined oxide and reduction plant at Listerhill, Alabama. Reynolds was a producer of aluminum foil for the cigarette industry, especially for the tobacco company of the same name. The tremendous capacity built for the government during the war, managed primarily by Alcoa, was sold to Reynolds and Kaiser after a Supreme Court decision in 1945.

Alcoa has its headquarters in Pittsburgh. Its first two factories were established in Pittsburgh and its suburb New Kensington at the end of last century when this region had access to very cheap energy in the form of local natural gas. Also the energy demanding glass industry was concentrated in this region as well as one of the electric industry giants, Westinghouse.

The three other aluminum leaders are Alcan, which is Canada based, Péchiney, which is part of the French metal-aluminum-group Péchiney Ugine Kuhlmann (PUK), and Alusuisse with headquarters in Switzerland.

The Canada-based Alcan (Aluminum Company of Canada) can trace its origin to the American antitrust legislation. Germany in the interwar period dominated production and consumption of aluminum by way of a state company which was the instigator of an international aluminum cartel. As an American corporation Alcoa could not participate directly but in 1928 Aluminum Limited (Alted) was formed. This company took over Alcoa's foreign investments except the bauxite mines in Surinam. Alted owned bauxite mines in France, Romania and Yugoslavia (Istria). Alcoa's large investments in Canada were concentrated into a subsidiary of Alted, Alcan, a company which still has large American interests.

Europe

Until the Second World War Europe dominated the aluminum industry. The bauxite deposits of Hungary and France were the leading raw material base. Oxide works were concentrated on the German brown coal fields and smelters utilized the cheap water power of the Alps (Austria, Switzerland, Bavaria, France and Italy) and the west coast of Norway. Ranked according to aluminum consumption at the end of the 1930s Germany was followed by the United States, Great Britain, the Soviet Union, France, Italy and Japan. These countries had a large aluminum production and as a rule net imports of aluminum. Only France and Italy were net exporters. Three countries rich in water power but with a small population had a large production that primarily was exported: Canada, Norway and Switzerland. Canada was surpassed only by Germany and the United States as a producer.

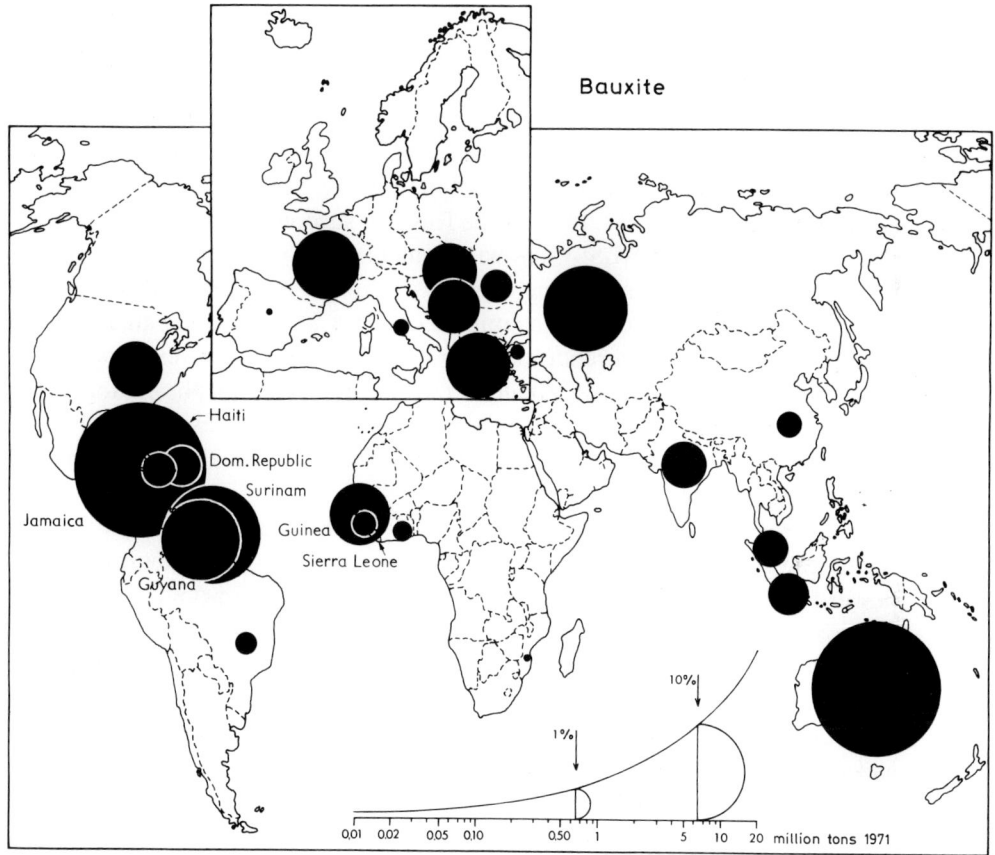

Fig. 35 — The global map of bauxite production has changed dramatically in the 20th century and is now dominated by tropical areas. In recent years, Australia has had a conspicuous expansion. Shipments from, above all, the tropical northern parts of the country in 1972 carried Australia to the top position among the bauxite countries.

The French bauxite mines are in Provence and Languedoc. The pioneer mine at Les Baux had been closed down for half a century, but was reactivated about 1970. The dominating French deposits at Brignoles in the departement Var near the Riviera has long accounted for 80% of the output but is expected to be phased down in the 1970s as a result of competition from rich deposits overseas, primarily in Australia. The Italian mines are along the Adriatic Sea, primarily at

Foggia and Bari. After the Second World War, Italy lost the mines in Istrija. Greece has bauxite mines in the Parnassos Mountains at Delphi and is one of the leading producers in Europe. Yugoslavia's mines are near Pula and Trieste, near the Dalmatian ports of Sibenik and Split, and at Niksic near the Albanian border. The Hungarian bauxite is mined in the Bakony Forest southwest of Budapest. Hungary has no cheap electricity and is a major supplier of bauxite to alumina works in East Germany and Czechoslovakia.

USSR

The Soviet Union was dependent on imported aluminum until the first alumina works with adjacent smelter was built at Volkhov east of Leningrad in 1932 and at Zaporozhye in 1933. In 1954 the Russians succeeded in obtaining alumina from nepheline ($KNa_3Al_4Si_4O_{16}$), with cement and soda as byproducts, which made the large apatite-nepheline deposits at Kirovsk in the Kola Peninsula the raw material base of a regional aluminum complex. It was already an important base of the Soviet fertilizer industry (phosphate). Nepheline moves by rail to the oxide works at Volkhov and Pikalovo (also located south of the Ladoga) and alumina as return freight to smelters at Nadvoitsy and Kandalaksha. The Northwest Region has a bauxite deposit at Plesetsk which supplies the alumina works at Boksitogorsk southeast of Tikhvin completed in 1938.

From the end of the 1950s, the aluminum industry was increasingly based on cheap water power from the Volga and the Siberian rivers and on thermoelectric power from the coalfields in Kuzbass. Increasingly alumina works and smelters were located in widely separate places. The large alumina works in the northern Urals, at Kamensk-Uralski and Krasnoturinsk which are based on the largest bauxite deposits in the Soviet Union (Krasnaja Chapochka) are integrated with smelters but also send alumina to Krasnojarsk, Bratsk, and Irkutsk in Siberia and to Volgograd, Sumgait, and Jerevan. Railroad distances are 2 500 – 3 000 km, in some instance even longer.

In the early 1960s it became obvious that the aluminaworks in the Urals could not keep pace with the expansion, primarily at the water power-based smelters in Siberia and the coal-based smelter at Novokuznetsk. New alumina works were established at Kirovobad, Achinsk, Rasdan and Pavlodar. The latter works is located in northern Kazakhstan on a southerly tributary of the Ob and receives its bauxite from Krasnooktjabrski och Arkalyk. Its alumina primarily moves to Novokuznetsk. Rasdan and Kirovobad exploit regional deposits of nepheline and send alumina to Sumgait and Jerevan. Achinsk is located where the Transsiberian Railroad crosses an easterly tributary of the Ob.

The shortage of rich aluminum ore in the Soviet Union has led to long-term import contracts with Yugoslavia and Greece and to a Soviet engagement in Guinea. The French (Péchiney) are going to build an alumina works near

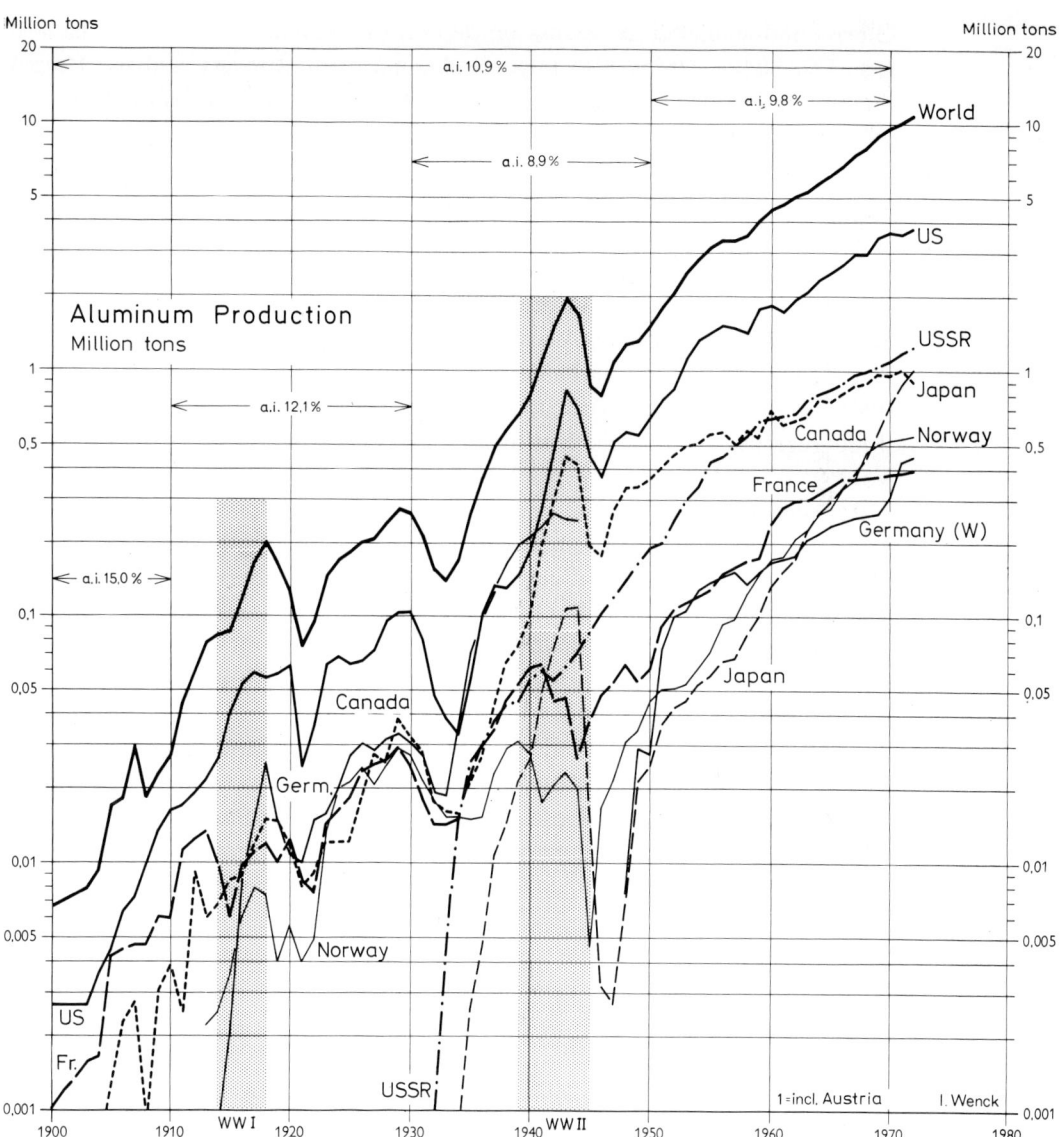

Fig. 36 — The exceptionally rapid growth of aluminum production went parallell with a gradual decline in the real price for the metal. Wherever the lightness of the construction material is of importance, primarily within the transport sector and especially for aircraft, aluminum plays an important part. Aluminum meets competition in the aerospace industry from titanium, which has a higher density but also a greater strength and therefore can be used in more slender structures with a lower total weight. So far, the high price of titanium is a serious restraint on its use.

Odessa and a smelter at Saiansk on the Yenisey River in exchange for natural gas. The Soviet Union also imports alumina from Hungary and the United States. The former moves to a smelter in Volgograd and the finished metal is sent back to Hungary.

North America

United States has large bauxite mines in Saline County southwest of Little Rock, Arkansas, but they account for only 4% of world production. The American share of the world alumina output is almost $1/3$ and of aluminum more than $1/3$. American and Canadian alumina works and smelters are located in three regions: the Pacific Northwest, the Southeast, and the St. Lawrence Valley.

Between 1902 and 1938 United States had only one alumina works. The East St. Louis plant had low assembly costs for the raw materials: bauxite from Arkansas, coal from southern Illinois and caustic soda. Smelters were built at water power plants at Niagara Falls (1895, shut down in 1949), at Shawinigan Falls in a tributary of St. Lawrence near St. Maurice in Canada 1901, at Massena on the New York side of St. Lawrence River 1903, at Alcoa, Tennessee 1914 and at Badin, North Carolina, 1916.

In 1926 Alcoa opened a reduction plant at Arvida on the Saguenay River in eastern Quebec and in 1928 an alumina plant. Arvida was the first integrated works in North America. The bauxite arrives by way of Port Alfred, primarily from Guyana. During the Second World War the giant power station at Shipshaw was built (1.3 million kW), somewhat upstream from Arvida. Before that, electricity was obtained from the smaller power station Chute-a-Carron further upstream. Canada got two more smelters at Beauharnois upstream from Montreal and at La Tuque near St. Maurice.

The late 1930s saw an expansion also in the United States. Alcoa in 1938 opened a second American alumina plant at Mobile, Alabama, designed for imported bauxite from Surinam. But it was during the war that United States surged ahead and became the leading aluminum producer in the world. To meet the demand for aluminum for the airplane industry, which rapidly became the leading manufacturing industry of the country, production was increased from 149 000 ton in 1939 to 837 000 ton in the record year 1943.

The expansion in the Pacific Northwest was remarkable. Smelters were built at Vancouver, Longview, Spokane, and Tacoma in Washington and Troutdale in Oregon. This region has the largest water power potential in the United States. Reclamation projects in the Columbia River basin were started in the 1930s, including the Grand Coulee Dam which houses a hydroelectric power station, long the largest in the world.

From zero in the beginning of 1940 the Pacific Northwest increased its share of the national aluminum production to 40% in 1944. In the postwar years the region even reached over 50%. On the Canadian side of the border the Kitimat smelter in British Columbia was completed in 1954. Nechako, a tributary of the Fraser River, was dammed and water from its reservoir on the eastern side of the coastal range was made to flow in a 16 km tunnel through the mountain range and to drop almost 800 m to a hydroelectric power plant on the coastal plain.

The location in the Northwest leads to record long railroad transports. Alumina moves some 4 000 km by rail from the Gulf Coast, where the alumina plants process bauxite from Jamaica, Surinam and Guyana, to Washington and Oregon. Much of the aluminum from the smelters move on by rail another 4 000 km to Chicago and other points in the Manufacturing Belt in the northeast.

Until the beginning of the 1950s all reduction plants in North America were based on water power. The first smelter built at a power plant burning natural gas was erected at Point Comfort, Texas. The Southeast, including the Gulf Coast, the Ohio River Valley and the bauxite fields in Arkansas, has expanded strongly in the 1950s and 1960s. Thermo-electric power plants have become more competitive in relation to hydro-electric power and improved transmission techniques have increased the value of electricity at hydro power sites, which all has helped reduce the competitive edge of a smelter location adjacent to water-power.

Since the Southeast has a favorable market location, many companies found this region with its good access to coal, oil and natural gas an interesting alternativ to the Pacific Northwest. One smelter even was based on brown-coal-generated electricity (Rockdale, Texas), common in Germany but unheard of in the Western Hemisphere. The Southeast was the largest center of aluminum production before the Second World War when the Tennessee Valley Authority (TVA) with its large supply of water power attracted aluminum smelters. But TVA has long had other large energy customers, e. g. the Atomic Energy Commission and has had to look for other sources of energy. The additions come from coal-fired thermo-electric plants and now also from nuclear reactors.

Latin America

The modern bauxite industry is dominated by large open-pits in the Tropics. Surinam and Guyana became leading producers in the interwar years. Mining occurs in the northern rim of the Guyana Shield. Since Surinam was a Dutch and Guyana a British colony the trade flows from Paramaribo and Georgetown were somewhat different. Both shipped bauxite to the United States and Georgetown in addition to Canada and Britain. In 1952 Jamaica became a bauxite producer,

172 Metals and Other Minerals

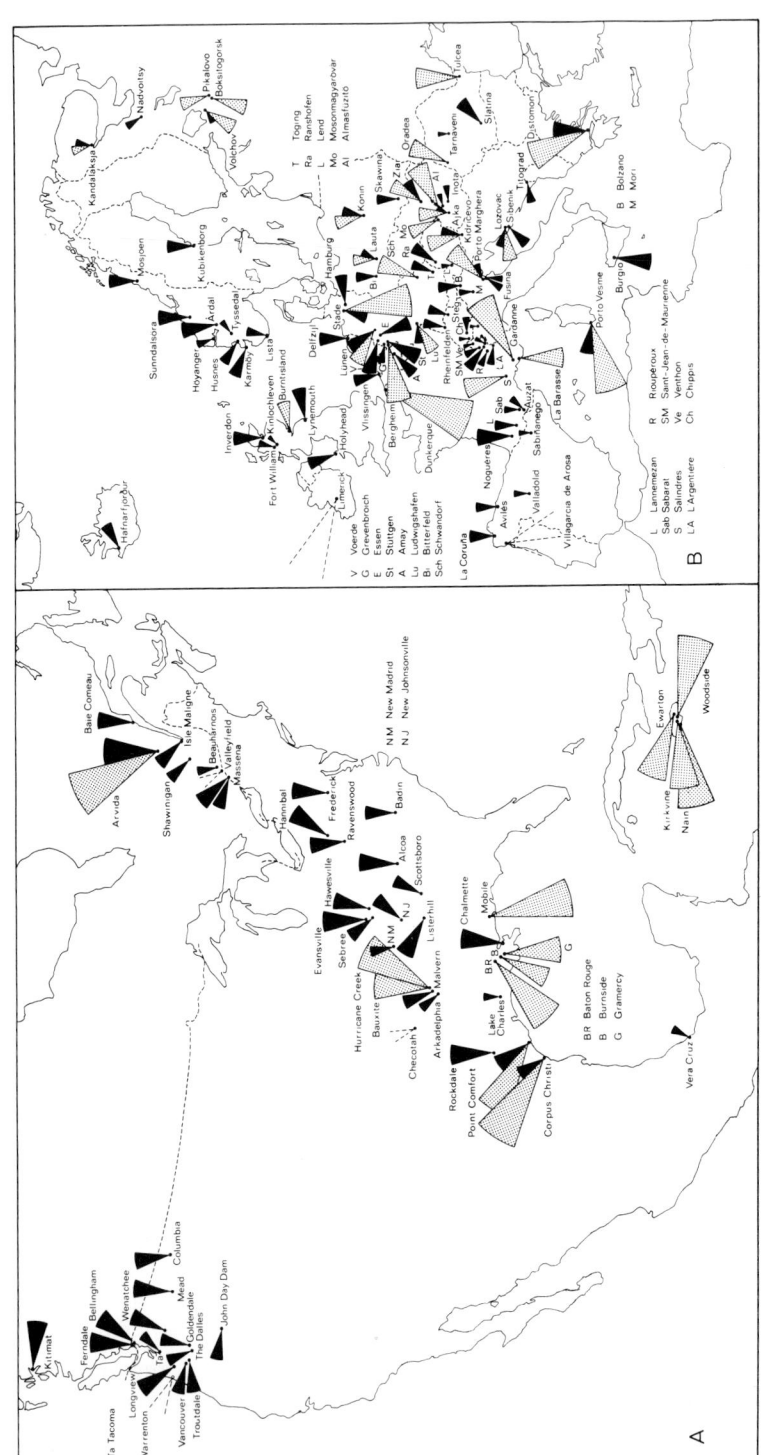

Fig. 37a

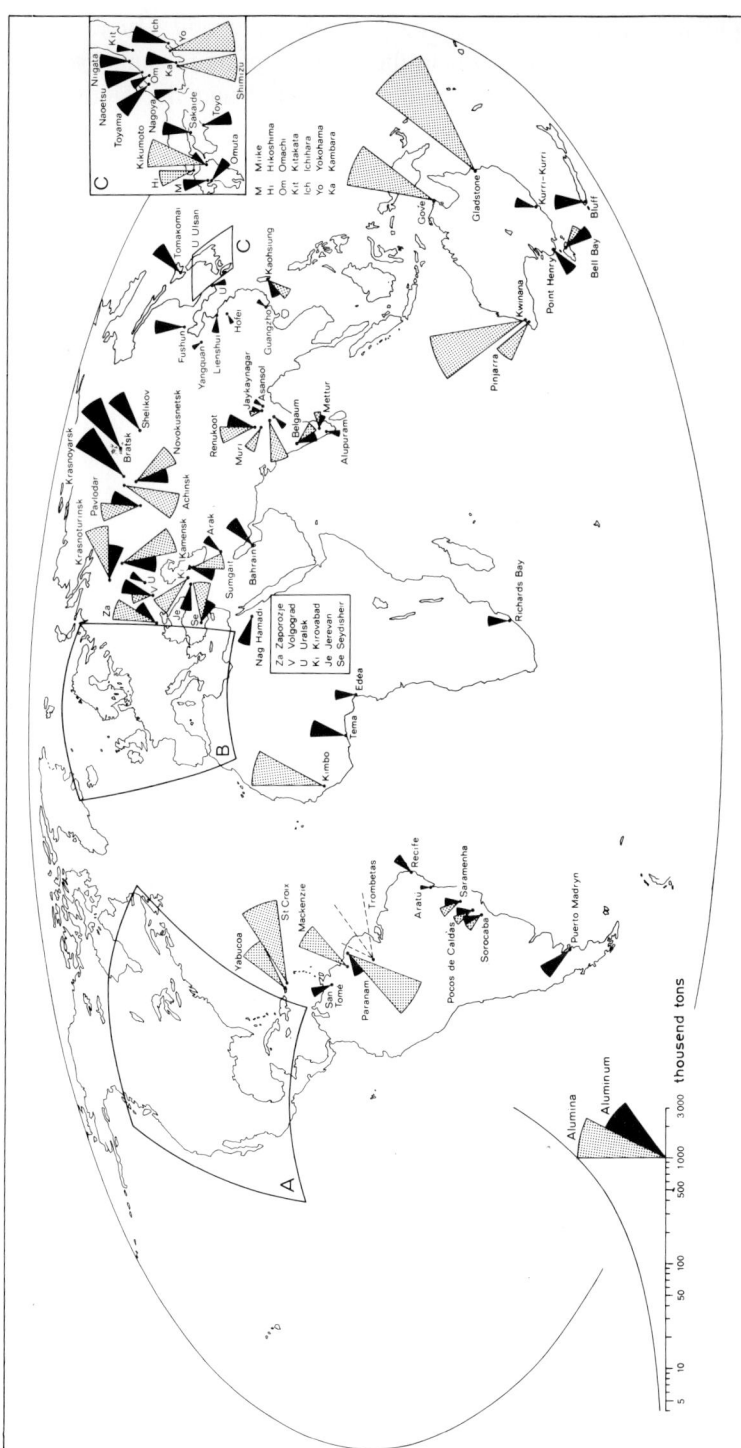

Fig. 37a – Fig. 37b – Alumina and aluminum works in the world 1975. Individual plants are shown with symbols, whose surfaces are proportionate to their capacity, see symbol scale.

primarily supplying alumina works in the United States and Canada, and Jamaica soon was the leading producer in the world, a position challenged by Australia in the early 1970s. In Hispaniola both Haiti and the Dominican Republic are bauxite producers. In the latter country the mines are in Sierra de Bahoruco.

The next large producer of bauxite in Latin America is expected to be Brazil in 1977. Multinational corporations (Alcan, Alcoa, Daniel Ludwig, Péchiney, the Shell subsidiary Billiton, a. o.) in cooperation with domestic interests have been prospecting since the middle of the 1960s at the Trombetas River, a northern tributary of the Amazonas. The river is accessible for ocean-going vessels. Both alumina plant and smelter are being planned.

Africa and Asia

The largest bauxite producer in Africa is Guinea with mines near Conakry and in the interior. The Soviet Union is involved in the development of one bauxite project in Guinea. Ghana and Sierra Leone are smaller producers. In Asia, India has bauxite mines in Madhya Pradesh, Bihar and Bombay and China has some production of bauxite and alunite. Bauxite from Johore in West Malaysia and from the Riow Islands south of Singapore, which belong to Indonesia were shipped to Japan already in the interwar years.

In the 1960s bauxite was discovered in northern Sumatra and a consortium of Japanese and American companies plan to build a hydroelectric power station in the Asahan River and a large aluminum reduction plant (250 000 t/y).

Japan has had a rapid expansion of its postwar aluminum production, the highest growth rate among the major producers. Production started in 1933 and in 1943 Japan reached a peak of 150 000 tons to meet wartime demands. Reduction works are scattered widely over the island of Honshu, initially oriented to hydro-electric power stations.

Australia

Tremendous bauxite deposits were discovered in 1955 at Weipa on the west coast of the Cape York Peninsula in Australia, which greatly changed the raw material situation in this part of the world. The mineral is exported as bauxite or alumina. The alumina plant, the largest in the world, is located in a different part of Queensland, at Gladstone south of Rockhampton. It is owned jointly by three competitors: Kaiser, Alcan and Péchiney, and two Australian groups: Comalco and CRA. Comalco, owned jointly by Kaiser and CRA, operates the mine which in 1972 had a capacity of 10.5 million tons a year. The first shipment from Weipa occurred in 1963.

Another large bauxite deposit has been developed by Alusuisse and domestic interests in the Gove Peninsula within the Arnhem Land's Aboriginal Reserve in the Northern Territory. Here the alumina plant is located at the shipping port (Dundas Point). Alcoa and domestic interests since the early 1960s exploit a deposit at the Darling Range east of Perth, long thought to be too poor for commercial use. The ore is shipped 45 km by rail from Jarrahdale to the alumina plant at Kwinana, an industrial suburb of Perth.

Magnesium

Magnesium (Mg) is the lightest of the structural metals. Magnesium has the density 1.74 (aluminum 2.70, titanium 4.5, iron 7.86, and copper 8.92). It is strong and machinable; it can be cast, rolled, drawn, spun, forged, blanked, and coined. Magnesium in alloys makes them corrosion resistant. But the high price precludes a wider use. Magnesium burns with a dazzling white light and is used in flares, fireworks and flash bulbs.

Magnesium can be extracted from underground brines, from sea water and from minerals containing the metal, primarily brucite, dolomite, and magnesite. Sea water is by far the most important source.

Sir Humphry Davy, the English chemist, isolated magnesium for the first time in 1808. But the element remained a laboratory curiosity until 1886 when a small plant for the electrolysis of carnallite ($KMgCl_2 6H_2O$) was built near Bremen. It was followed by a full scale plant at Bitterfeld. The firm, Griesheim Elektron, was later merged into a chemical company that in 1925 participated in the large chemical trust, IG Farbenindustrie.

In the United States production was started in 1915 when the German deliveries stopped. Dow Chemical Co. produced magnesium from a subterranean brine at Midland, Michigan. Research and development by Dow in 1928 led to an electrolytic cell that in the Second World War accounted for over 90% of the American magnesium production. Since magnesium is an alloy metal of aluminum (airplane industry) and is used in bombs and other munition, the American production reached a pronounced peak during the Second World War but also the Korean War and the Vietnam War led to increased production.

To meet the increased wartime demand the United States built three new plants at Freeport and Velasco, Texas, based on sea water. Also other raw

Tab. 46 → The American Magnesium Production in Selected Years
(thousand metric tons)

1918	1939	1942	1943	1946	1953	1962	1965	1970	1972
0.13	3.04	44.42	166.55	4.82	84.44	62.56	73.80	101.60	109.60

materials for the chemical industry were abundant here: oyster shells, salt and sulphur. Not the least important was the easy access to cheap natural gas for the production of electric energy. The largest wartime plant was built at Las Vegas, Nevada. Its capacity equalled that of the three sea plants combined. It has not operated in peacetime.

Tab. 47 – World Production of Magnesium 1970
(thousand metric tons)

United States	102	Japan	10
USSR	50	Canada	10
Norway	35	Italy	8

| Total | 223 | | |

Source: MY 1970, Vol. 1.

Titanium

Titanium (Ti) has been known as a common element for more than 150 years. It occurs in almost all rocks and in two important ores, rutile or titanium dioxide (TiO_2) and ilmenite ($FeTiO_3$). Ilmenite often occurs in monazite sand. Ferro titanium is used to bind carbon and nitrogen in special steel qualities. Titanium dioxide or titanium white is a waterinsoluble powder used in white pigments, plastics, ceramics and for delustering synthetic fibers. It is also used as a coating on welding electrodes. These uses account for the largest quantities of titanium ore produced.

Although titanium is one of the ten most common elements in the earth's solid crust, table 2, the pure metal is still expensive, some US $6.60 a kilogram in 1970. Production per ton requires more than twice as much energy as aluminum and some steel alloys. The metal, which is obtained from rutile, is only used where lightness, strength (especially at high temperatures, 600°C) or corrosion resistance justify the cost. Over $4/5$ of the titanium metal is used in bracing constructions of airplanes, in ramjet and turbo jet engines and in rockets. Production has followed a steeply rising curve since 1948. Titanium and its alloys have tensile strengths comparable with those of many steel alloys and their corrosion resistance at normal temperature corresponds to that of platina. Pipes in desalting plants is a new use of titanium metal. Its corrosion resistance to salt water is unique.

In Australia ilmenite is obtained as beach sand in southern West Australia and rutile along the coast between Sydney and Brisbane. The sand mineral production is dominated by some multinational corporations: Consolidated Goldfields, Conzinc Riotinto and others. Also in the United States, Malaysia,

Tab. 48 – World Production of Ilmenite and Rutile 1970
(thousand tons)

Ilmenite:

Australia	887	Norway	579	Sri Lanka	82
United States	787	Malaysia	192	India	79
Canada	766	Finland	151	Spain	44

Total	3575

Rutile:

Australia	368	Sierra Leone	44

Total	417*

* Reliable data for the Soviet Union are missing. A titanium mine exists at Kusa in the central Urals.
Source: MY 1970, Vol. 1.

India and Sri Lanka ilmenite is produced from placers. American sand quarries are primarily located in the southeastern states including New Jersey. Allard Lake, Quebec, is the largest known titanium ore body, a layered intrusion like the Sanford Lake deposit in the Adirondack Mountains of upstate New York.

In Norway and Finland ilmenite-magnetite is produced in mines. Finland has a large mine at Otanmäki south of Oulujärvi. The Norwegian mine at Sokndal between Egersund and Flekkefjord is among the largest titanium deposits in the world. The large ore body in Smålands Taberg, Sweden, has been mined only for its iron ore. The Russian deposit in the Ilmen Range of the Urals was the namer of ilmenite.

Nonmetallic Minerals

Nonmetallic minerals other than fuels and fertilizers cover a wide range. Many minerals used in the construction, ceramics and chemical industries are abundant and cheap but since total quantities used are very large the combined value of cement, clay, gravel, lime, sand, slate, stone, gypsum, salt and sulfur is larger than the value of iron ore, bauxite and several other major metal ores combined. Several minerals are ubiqities; if not literally found everywhere they are at least available in most regions of the world. They often play an important part in short distance transport on land and on the waterways but seldom loom large on long distance routes.

Several other nonmetallic minerals are important in our civilization although used in small quantities: asbestos, graphite, mica, quartz crystals, diamonds and gems. The electronic industry is a heavy user of quartz crystals.

Stone

Stone is widely used in all parts of the world but as a house building material it was relatively more important in the past, especially in densely populated regions with restricted forest areas such as the Mediterranean countries, France and Britain. The laborintensive stone-cutting industry is now hardly in a position to compete with manufactured stones, such as bricks and cement blocks, and with reinforced concrete.

Only a few per cent of the stone used is cut into dimension stone; the rest is sold as crushed stone. The more important stones utilized by man are basalt, granite, limestone, marble, sandstone and slate. Almost all dimension stone and some 75% of the crushed stone are used for construction. Another 15% of the crushed stone is turned into cement which is also used in construction. Limestone and dolomite account for $^3/_4$ of the total tonnage.

Most dimension stones are used relatively near the quarry but some building stones, because of exceptional beauty or special properties, are shipped long distances. There are several times as many crushed-stone plants as quarries producing dimension stone. The local geology largely decides which type of stone that is used a given place. Formerly this was even more pronounced. Old settlements in regions with houses predominantly of stone clearly reflect the local geology. Some $^3/_4$ of the crushed stone is hauled by truck, which indicates a short average haul.

Basalt and other dark igneous rocks, known commercially as traprock, are used almost exclusively as crushed stone. Granite, an igneous rock that varies much in color and crystal structure, takes a high polish and is used in monuments and tombstones. Barre, Vermont and Quincy, Massachusetts, are well-known for their granite. Only 2% of the American granite is cut into dimension stone but this represents $^1/_3$ of the value. Like traprock, crushed granite is used as concrete aggregate, roadstone and railroad ballast. Granite as a paving stone in towns can no longer compete with asphalt and other materials. In Europe, Sweden, Finland, Norway and Scotland are known for their dimension stone. Bohuslän and Blekinge in Sweden used to export paving stone from coastal quarries.

Limestone is second only to granite as a dimension stone in the United States. The Bedford limestone of the Bedford-Bloomington area in Indiana accounts for over half the tonnage. Its uniform gray or buff color, pleasing even after weathering, makes it competitive in distant markets. The Lannon limestone of Wisconsin is also much in demand.

The dominant position of limestone expressed in quarried tonnage results from its use as flux in steel mills and as the raw material of cement and lime. Cement and lime factories are located close to the limestone quarries. In the assembly costs for raw materials of a steel mill limestone ranks after iron ore and coking coal. But few blast furnaces are far away from a limestone quarry. For coastal mills, cheap sea transport extends the economic distance that limestone can travel. On the Great Lakes and in the Baltic, coastal quarries play an important part. The world's largest limestone crushing plant is located at Rogers City, Michigan. It ships some 15 million tons a year. Limestone ranks among the major products shipped on the Lakes and in the Baltic.

Marble or metamorphic limestone takes a high polish and is used for tombstones, monuments, statuary and buildings. Proximity to navigable water and to population centers influence the competitiveness of marble quarries as well as other stone production but their high prices allow some types of marble to be shipped long distances. Major producing centers in the United States are Vermont, primarily the west flank of the Green Mountains, Knoxville in Tennessee and northern Georgia. Carrara, Italy, on the west flank of the Apennines near the Gulf of Genoa has been mined for centuries for its snow-white, fine-grained marble used for much of the world's classic sculpture.

Slate or metamorphic clay or shale has well-defined planes of cleavage and is excellent for roofing, for flagstones, steps, sills and so on. Quarries in northern Wales claim to have been in production since the reign of Elizabeth I. In the United States Pennsylvania, Vermont and New York have the largest production.

Sandstone like slate and limestone is used for flagstones and steps but is less popular for construction than formerly. The Amherst quarry near Cleveland, Ohio, known as the Gray Canyon, is the world's largest sandstone quarry. In addition to Ohio, Pennsylvania, Tennessee and Maryland have large production.

Cement

World production of cement in 1972 was some 640 million tons. After an annual increase of 7.4% since 1950, a little more in the 1950s (9.1%) then in the 1960s (6.0%), the cement tonnage for the first time exceeded that of steel in 1971. Cement is used almost wholly by the building and construction industry. Mixed with water and an aggregate of sand, slag, gravel, or iron rods, cement forms an artificial stone known as concrete.[23]

A patent for portland cement was taken out in 1824 by a bricklayer of Leeds, Joseph Aspdin. It was named for the British Isle of Portland where a Jurassic limestone of similar appearance, used in London as a building stone, was quarried. It took time for portland cement to be improved and accepted. Its competitors were natural cement, or the burned and pulverized cement rock, and puzzolan, a mixture of slaked lime and granulated blast furnace slag. Until 1900 natural cement made up over 60% of the American cement production. Portland cement, which hardens even when kept under water, is also referred to as hydraulic cement. The uniformity of cement according to rigid specifications was ensured in the early years of this century.

Technique. Raw materials are mixed and ground and fed into the upper end of an inclined rotary furnace or kiln. From the lower end comes a flame fed by pulverized coal, oil or gas. The material is fused to clinker at a temperature of 1450°C. Two processes are used for feeding raw materials into the kiln. In the dry process, used with dry materials such as cement rock and limestone, the material passes through rotary driers. The more common wet process is employed when raw materials have a high moisture content. The material is crushed and mixed with water to form a thin mud or slurry. The resulting clinker is a calcium aluminosilicate which contains 64% calcium oxide, 5.5% aluminum oxide, 21% silicon oxide, 4.5% ferric oxide, 2.4% magnesium oxide, and 1.6% sulfate. In some places the quarried rock is of the desired composition but as a rule combinations of limestone, chalk, oyster shells, marl, clay, shale, iron ore, slag and silica sand have to be mixed. Some 75–80% limestone, or its equivalents, and 20–25% clay or shale are common mixtures. Kilns vary greatly in size, the largest having capacities over one million tons a year, but large cement plants often have a battery of kilns.

Before 1900 clinker was produced in vertical shaft kilns, which had to be unloaded and charged between two burnings. They required much labor but were economical in the use of coal. The rotary kiln, first used at Ålborg[24], Denmark, and at Northampton in the Lehigh Valley of eastern Pennsylvania at the turn of the century, operates continously, uses more fuel but is much less labor intensive.

To the cooled clinker is added 2% gypsum to improve the setting qualities and the mixture is ground into a fine powder which is shipped in bags or in bulk. A modern cement mill uses heavy equipment in the quarry, in the crushing and grinding departments, in the kiln department, and in the finish and shipping departments. A continuing trend is for larger units of equipment. Computer control, covering most operations, is common. For every ton of portland cement shipped, over $2\frac{1}{2}$ tons of raw materials and cement clinker must be ground to the fineness of flour; $\frac{1}{2}$ ton of coal or equivalent fuel is burned.

The dusty emissions of the cement factories have caused serious pollution problems. In the postwar period large amounts have been spent on pollution abatement systems. Some 20% of the capital investments in new cement plants were for such facilities in the early 1970s. The emission of fine lime powder, which used to give a grey color to everything in the neighborhood of a cement plant, has been reduced to insignificance. Other, less visible forms of air and water pollution, were given attention and so was noise pollution. Noise from grinding mills was muffled and control rooms were soundproofed to protect operators.

Location. Cement is a low-priced mass product. Economies of scale are obvious in all steps of production and transport. The position of raw materials and markets greatly influence the location of cement mills. Fuel accounts for a considerable portion of production costs, but price differences for delivered fuel at hypothetical points in the landscape are small. A location at a limestone quarry near a major market, and preferably on a waterway, is common.

In the early years, England, Germany and Belgium had a large share of the cement industry, which is partly explained by their early start but primarily by their freight advantage. For obvious reasons, more cargo has always been shipped into large industrial agglomerations than leave these agglomerations in the form of manufactured products. In the days of sailing vessels, many ships left London, Rotterdam or Antwerp in ballast, usually sand, rocks or soil, which was unloaded in the harbor of destination. Instead of ballast, the ships could carry cement bags, if available in the estuary for the right destination. The lower Thames River between Dartford and Gravesend got a large concentration of cement mills that used an abundant supply of excellent chalk from the North Downs and clay dredged from the river. Rivers and canals in the hinterlands of Rotterdam and Antwerp were attractive locations for export-oriented cement mills.

One of the largest cement mills in Europe, at Amöneburg near Wiesbaden on the Rhine, sent cement cheaply downriver by barge to Rotterdam in transit for New York or New Orleans where it could be delivered cheaper than in Berlin or Munich. European cement was highly competitive in America in the days of the labor-intensive vertical kilns and sail ships. The freight advantage remained on a reduced scale when steamers took over more and more of the cargo flow but the capital intensive rotary kiln made American cement more competitive on the domestic market. In the Southern Hemisphere and in the Tropics, European cement had the advantage not only of a return freight position but of a slack in domestic demand during the winter months when construction reached a seasonal low. Construction workers were laid off in the cold season, but the cement mill had to work continously to be profitable. In the late 1920s, Belgium alone accounted for some 30% of all cement that entered international trade.

In the United States, the Lehigh Valley of eastern Pennsylvania had the advantage of an early start and excellent limestone deposits, a central location in the large Megalopolis market and a return freight position by rail vis-à-vis the rest of the country. It produced some 70% of the American cement at the turn of the century. In 1970, cement was produced at 181 plants in 41 states and Puerto Rico; eastern Pennsylvania accounted for only 8% of the American output. Texas, southern California and Michigan equalled or surpassed eastern Pennsylvania in cement production. The American development epitomizes the changed distribution pattern in the world. The construction of taller buildings, the building of highways, especially bridges and overpasses, the many new

182 Metals and Other Minerals

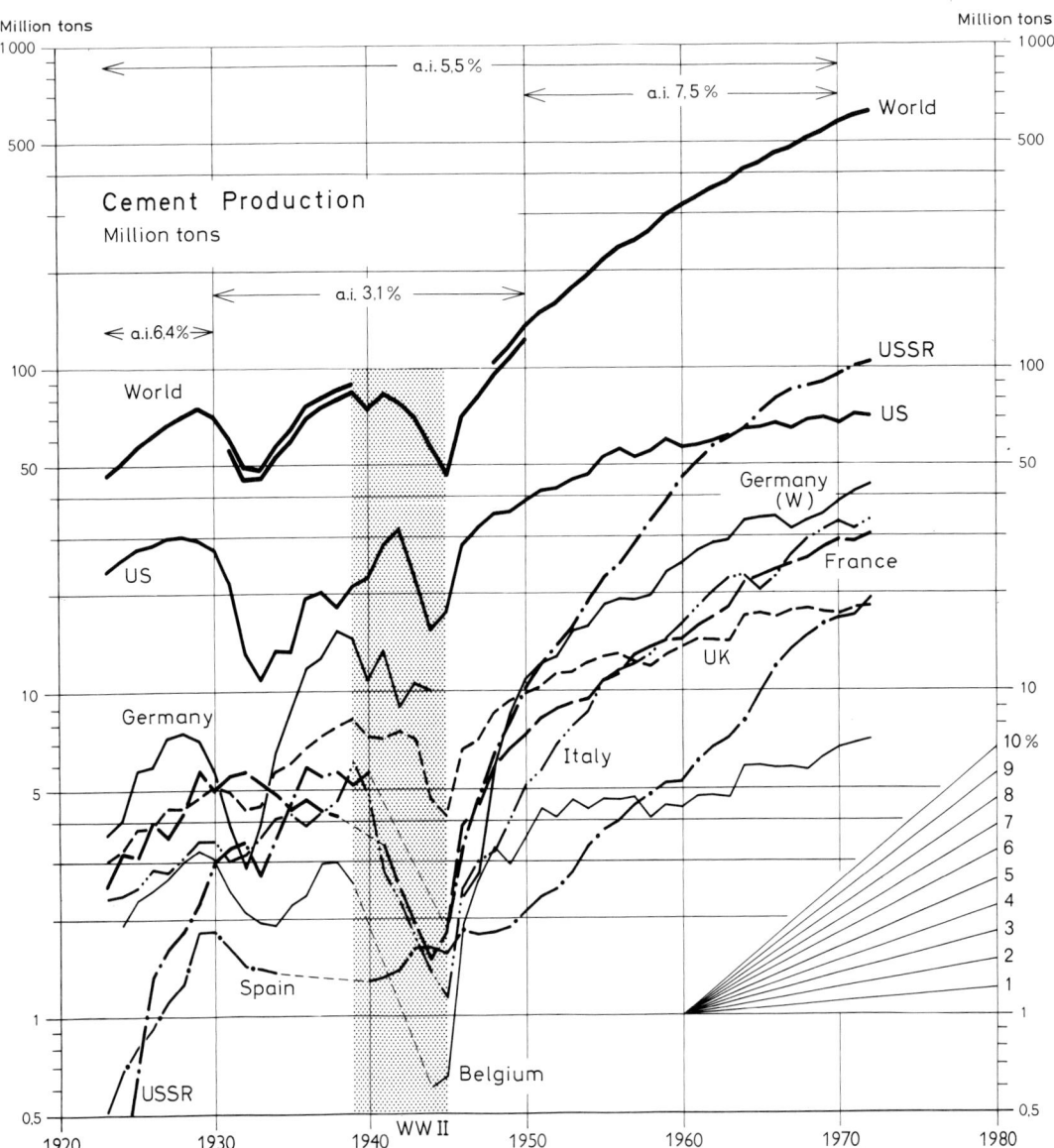

Fig. 38 – Cement production 1923–1972. United States, a dominating producer in the 1920s with four times the production of Germany, had a slow rate of increase in the post-World-War II period. In 1973, it was in third place after the Soviet Union and Japan (not shown): USSR 110, Japan 78, and US 78. Belgium competing with Italy among the major cement producers in the 1920s and a leading exporter to the world market, has fallen far behind. See text for explanation.

airfields, expansion of harbors, the construction of dams and power plants, all have contributed to the rapid increase in the market for cement that justifies the establishment of cement plants in many small nations and regions. With a more even distribution of plants, interregional and international trade in cement becomes rather small, only about 3% of world production crosses national boundaries. Some countries and regions may temporarily run into overcapacity while others have undercapacity. The industrial countries still have a freight advantage over the underdeveloped countries. Japan, Belgium, France, Germany, Norway and Italy in recent years have been leading exporters of cement to the world market.

The cement industry and economic theory. The manufacture of cement is a process industry in which the economies of scale are obvious. Both raw materials

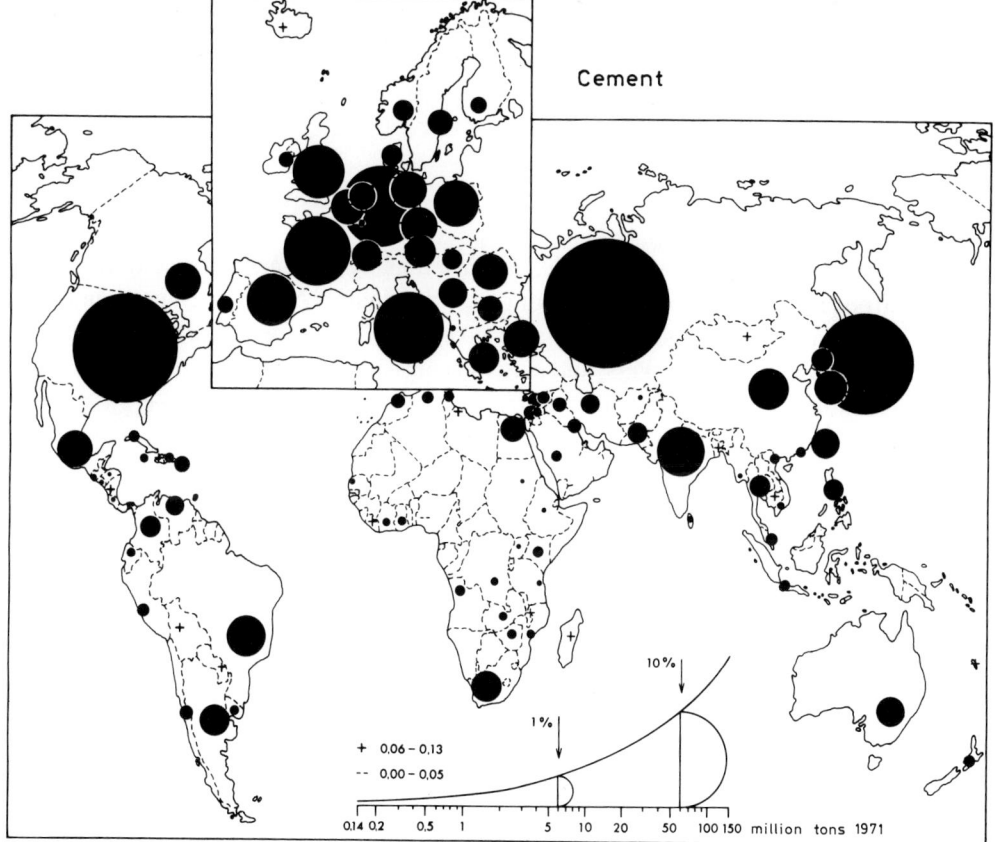

Fig. 39 – In the ten-year period before 1973, Japan and Spain were the most rapidly expanding cement producers, partly working for an export market. The cement industry of the 1970s is one of the most widely spread mineral industries. Almost all countries and regions have cement factories.

and product are lowpriced. Since raw materials are more or less ubiquitous, the market is the important location factor. As in the steel industry, cement plants should have as large a home market as possible; absorption of freight for sales in the markets of other cement plants should be at a minimum.

The history of cement plant location offers interesting examples of external economies. A. Marshall (1896) used the apparel industry of London as his archetype of factories going outside their own premises for savings on unit costs. He could as well have used the cement plants on the lower Thames. They should, theoretically, have been paid by the harbor authorities when they dredged clay from the river bottom (and deepened the approaches to the harbor). Since cement bags were more handy as ballast in sailing vessels than soil or sand, the conventional ballast, ship owners might, theoretically, have been asked to pay for carrying cement from London to overseas destinations. Cheap raw materials and low distribution costs for the finished product were external economies available only to cement plants in or within easy access to the large ports of north-western Europe, primarily London, which in those days was the focus par excellence of world trade.

Asbestos

Asbestos is a fibrous amphibole, used for making thousands of incombustible or fireproof articles. A few major uses dominate. Asbestos cement building materials and asbestos cement pipe take some 70% of world production. Floor tile, brake linings, gaskets, and clutch facings are other important uses.

Canada and the Soviet Union dominate world production, which was 3.5 million tons in 1970. Canada produces over 40%, but asbestos accounts for only 4% of the value of the Canadian mineral production. The vast deposits of asbestos in southeastern Quebec, in a narrow strip from the American border to the St. Lawrence River with Thetford Mines as the major producing center, were discovered in 1877 in connection with the construction of the Quebec Central Railway. For many years these deposits were exploited in open pits, but most operations are now underground. Other asbestos deposits have been developed in northwestern Quebec near Amos and Val d'Or, in Newfoundland, Ontario and British Columbia. The Soviet output was long concentrated to the Urals, but in 1965 production started at Dzhetygara in northwestern Kazakhstan.

South Africa, China, Italy and the United States are relatively small producers of asbestos.

Clay

Clay is widespread and used for many purposes. It consists essentially of hydrated aluminium silicates. Common clay is used in tens of millions of tons. It is almost exclusively used by the producer in the fabrication of a product: common brick, tile, sewer pipe, drain tile, and cement. In the tropics and in dry areas in the middle latitudes, adobe in large but unrecorded quantities is mined and processed almost entirely by hand, very close to the place, where it is being used.

Sand and Gravel

Sand and gravel rank first in tonnage among the minerals, surpassing oil and coal. Both result from natural disintegration of rocks, and primarily consist of the harder minerals, such as quartz, which is silica or silicon dioxide. About 90% of all sand and gravel is utilized for construction and paving. Sea shores and sea bottoms, lake beaches and lake bottoms, and river terraces and river bottoms are the major sources of sand and gravel except in formerly glaciated areas where eskers and glacial outflows provide much material. In areas with no natural sand and gravel, rocks are crushed to provide substitutes. The sand and gravel industry is highly decentralized, each small unit operating within a restricted area. About 6000 commercial units accounted for 80% of an American output of over 800 million tons in 1970; the rest was produced by the government and by contractors. Trucks hauled 90% of the tonnage. The terms of the mining lease may now stipulate the creation of recreational lakes upon termination of the mining operation.

Tab. 49 – Soils by Particle Size
(diameter, mm)

Boulder	> 200	Medium sand	0.6 −0.2
Stone	200−20	Fine sand	0.2 −0.06
Coarse gravel	20− 6	Very fine sand	0.06 −0.02
Fine gravel	6− 2	Coarse silt	0.02 −0.006
Coarse sand	2− 0.6	Fine silt	0.006−0.002
		Clay	< 0.002

Glass

Glass was made in Antiquity by the Egyptians and Romans. As a handicraft, glassmaking was well developed centuries ago among the Venetians, Dutch and French. Mechanized glassmaking is new, however, developed since about 1900.

Almost all window panes and glass containers are now machine-made; a hundred years ago even the flat glass was handblown.

The raw materials for glass are sand, soda ash, lime and a large number of other materials in smaller quantities, i. e. metals or their oxides to give color to the glass (cobalt, blue; gold, red; copper, red or green, and so on). Sand, which accounts for the largest quantities, should have a low content of iron and aluminum oxide and a high silica content. Sand grains should be uniformly small and angular which simplifies the fusion. Broken glass or cullet shortens the fusion time and lowers the cost.

Melting takes place in a regenerative furnace, developed in the 1860s by William Siemens of the famous German industrial family but working in England and by the French metallurgist Pierre Martin. The same Siemens-Martin furnace has dominated the steel industry until recently (open-hearth furnace). This fuel-saving furnace allowed an increase in batches to over 1 000 tons. It became advantageous to be located near cheap gas, the ideal fuel, either by-product gas from coke ovens or natural gas. In the United States the industry was strongly concentrated in the Pittsburgh region, the first developed natural gas area, later with access also to by-product gas from coke ovens. In Europe the glass industry was often located near the coke ovens. It had one of its strongholds at Charleroi on the Sambre River in Belgium which has remained an important center of the plane glass industry with much export. Even before the modern natural gas era the industrial regions were crisscrossed by pipe lines for gas. The glass works did not have to be located adjacent to the coke ovens.

The glass industry is divided into three main divisions according to its products: flat glass, containers and special glass. The first two are strongly mechanized and have obvious economies of scale. In the special glass division development costs are often high. In the quantitatively small but culturally important art glass industry craft skills are decisive and wages make up an important cost item. Production units therefore as a rule are small.

The flat-glass industry underwent a revolutionary development at the beginning of this century with the introduction of the continuous, vertical drawing of flat glass from the surface of the glass tank through a series of rollers. The continuous sheet-drawing process was invented by the Belgian Fourcault in 1902 and was further developed in the United States by the Pittsburgh Plate Glass Company, now PPG Industries. The process was so cost saving that the whole flat-glass industry in America switched to the new method within 15 years before 1935. Belgium has contributed cost saving inventions even after the Second World War, e. g. a machine for grinding and polishing plate glass on both sides simultaneously.

The vertical drawing of flat glass has met competition from a method invented in 1959 at the largest British glass works, Pilkington Brothers at St. Helens near

Liverpool. Flat glass here flows continuously in a horizontal direction on a bed of molten tin. By 1974 over 50 glassworks in the world had switched to the float-glass method. PPG, who produces float-glass on a licence, in 1975 stated that the company had patented a new method for making flat glass that would yield a product of a more even thickness and better optical characteristics and at lower cost than the Pilkington-method.

The flat-glass industry is dominated by large corporations and large production units. In the United States three corporations account for three-fourths of the window glass and two of these, the Toledo-based Libbey-Owens-Ford and the Pittsburgh-based PPG Industries, for 90% of the plate glass. The PPG Industries have diversified and have more than half their production in other lines than glass, primarily paint which is being distributed from the same regional warehouses as window glass.

Glass containers — The first fully automated machine for bottle blowing was patented in 1899 by the American glass blower M. J. Owens. Mechanization led to a rapid productivity increase. At the turn of the century one man produced 40 000 containers a year and in 1955 no less than 400 000. In the same period the number of hours worked declined substantially. Lowered production costs led to lowered prices, increased demand and new markets.

Two American companies manufacture the complex machines, Owens-Illinois and Hartford-Empire. In 1924 the two companies agreed on a cross-licensing system which led to investigations by the antitrust authorities about restraining influences on competition in the glass industry. Competition among remaining companies in the industry is now thought to be satisfactory in spite of the concentration to a few groups. Owens-Illinois accounts for over one-third of the American production and uses its own machines.[25] The company produces over 5 000 types of containers and employs more than 69 000 people in many factories scattered from coast to coast. Like its competitors, Owens-Illinois must weigh its desire to be close to its competitors for fast delivery and good customer contact against the need to be large enough to justify the heavy investments in machinery. The modern distribution net for natural gas has reduced the importance of the gas fields for the location of the glass plant. The assembly costs of energy and bulk raw materials have long been lower than the distribution costs of the finished products. This is another way of stating that the glass industry, like the steel industry and other industries that, in transport terms, transform bulk cargo into general cargo, has become market oriented.

Special glass includes everything from Christmas decorations, pressed gift articles and table glass to crystal glass and other forms of art glass.

In recent decades, glass has been used in a number of new fields. Glass blocks for the construction industry, bulbs and photo flashes as well as picture tubes and picture screens for TV-sets are examples as is the glass fiber, introduced in 1938, which is used for boat hulls and in automobiles, as filter and for sound and heat insulation. In the electronic industry glass continously gets new uses. The best known glass company within many of these new fields is Corning Glass Works with plants in several states in addition to the main factory in Corning, New York. The company has a remarkably research- and development-heavy budget and employs 46 000 people.

Tab. 50 – International Trade in Glass 1971 (SITC 664)
(million US dollars)

Imports		Exports	
United States	123	Belgium	175
Germany (W)	96	United States	117
France	68	Germany (W)	115
Canada	64	France	78
Netherlands	50	United Kingdom	67
Italy	49	Italy	49
Total (excl CC)	863	Total (excl CC)	776

excl CC = communist countries excluded
Source: YITS 1972–73.

Tab. 51 – International Trade in Glassware 1971 (SITC 665)
(million US dollars)

Imports		Exports	
Germany (W)	80	Germany (W)	116
United States	74	France	116
Canada	50	United States	68
Netherlands	49	Italy	58
United Kingdom	38	United Kingdom	51
France	35	Belgium	45
Total (excl CC)	679	Total (excl CC)	628

excl CC = communist countries excluded
Source: YITS 1972–73.

Porcelain

Certain types of clay represent higher values per ton and sometimes are shipped long distances. Kaolin, named for Kaoling, a Chinese mountain which yielded the first kaolin sent to Europe, is produced in Cornwall and Devon in southwest

England and exported to chinaware works throughout the world, primarily through the adjacent port of Fowey. Kaolin is also used for paper filling.

The term china indicates that the product originally was imported from China; it was an important item in the early Canton trade. Kaolin is now mined in Kiansi and Anhwei provinces as well.

The modern porcelain industry was started by the Germans in the 18th century. Dresden, Meissen, and Berlin became well-known for their china. Austria, Czechoslovakia, and Denmark also have famous china factories based on nearby raw material. In France, the western flank of the Massif Central has fine kaolin and Limoges is known for its porcelain. In Britain the North Staffordshire coalfield is known as the Potteries. Here six pottery towns early in this century were merged to Stoke-on-Trent, which is now the center of the industry. Stoke gets its fine clay from Cornwall through the Trent and Mersey Canal, opened in 1770. Wedgwood china from this area is well-known.

In the United States, Georgia produces $3/4$ of the kaolin. The pottery industry in its early stage was strongly centered on Trenton, New Jersey, but from the end of last century the upper Ohio Valley has dominated. The district straddles the borders of three states: Ohio, Pennsylvania and West Virginia; East Liverpool, Ohio, has long been known as the American pottery center. Lennox and Syracuse, New York, also have pottery industries.

Ball clay

Ball clay or common blue clay is used for white ware, high-grade tile, and enameling. England and Germany are major producers. In the United States most is produced in Tennessee.

Fire clay

Fire clay is produced more widely. Five states dominate the American output: Ohio, Pennsylvania, Missouri, California, and Alabama.

Bentonite

Bentonite is used as a binder in forming strong pellets of iron ore. This demand is worldwide, and the American deposits are the principal source. Swelling bentonite, used for this purpose, is produced in Wyoming, Montana, and South Dakota. It is also used for rotary-drilling mud. Bentonite and barite are important ingredients of the 'mud' used in petroleum drilling. It is not unusual that the mud for one well costs more than a million dollars. In addition to cooling the drill steel it helps keeping the pressure in the hole and thereby prevents petroleum under pressure from blowing out.

Fuller's earth

Fuller's earth or 'bleaching clay' is a nonplastic clay, high in magnesia, used in filtering, decolorizing, and clarifying but also for absorbent and filler purposes. It is widely found in the world. In America it is produced along the western Georgia–Florida border between Attapulgus and Quincy.

Diatomite

Diatomite or diatomaceous earth is a chalklike, porous material composed of fossils of diatoms. It is primarily used in filtration of sugar, water, and beverages but also as a filler in rubber, paper, paints and so on. Commercial deposits are usually of Miocene age. United States accounts for $1/3$ of world production. Southern California (Lompoc and San Pedro) is the major district followed by other western states (Nevada, Washington, Arizona and Oregon). Other major producing countries are the Soviet Union, Denmark and France.

Graphite

Graphite, named for one of its minor uses, in 'lead pencils' (Gr. graphein, to write), is a crystalline form of carbon. It is used in foundries for molds and crucibles, in refractories, and for several other purposes.

Few of the many widespread deposits are large enough or sufficiently pure to be exploited. Ceylon and Madagascar have long been known for their vast deposits of high-grade crystalline graphite, but the largest producers in the world are Korea, North and South, and the Soviet Union. Mexico and Austria rank with Sri Lanka (Ceylon) and Madagascar among the major producers.

Sulfur

Sulfur (S) is primarily, over 85%, used for making sulfuric acid, the 'work horse' among the chemicals used by industry. The largest tonnages are required to turn phosphate into superphosphate. Half the weight of superphosphate consists of sulfuric acid. The fertilizer industry takes about 50% of all sulfur. Another 25% are consumed by the pulp and paper industry, for inorganic pigments, nonferrous ore leaching, and explosives.

In 1970 the global production of native sulfur by the Frasch process was 10 million tons and from sulfur ores 3 million tons. Another 9 million tons of elemental sulfur was recovered from sour petroleum. Various pyrites, including cupreous pyrites, supplied some 10 million tons of sulfur.

The Frasch process is used to obtain sulfur from the cap rock of salt domes found along the Gulf Coast of the United States, in Texas and Louisiana, and of Mexico, in the Isthmus of Tehuantepec. Similar salt domes are found in large numbers east of the lower Volga. Poland is the only other area where the Frasch process is being used. In 1970 Poland, now a major exporter of sulfur, helped Iraq initiate exploitation of her large known reserves of sulfur by this process. The unique mining system is based on sulfur's low melting point, 115–120 °C. Superheated water is forced down the outer of three pipes, one inside the other, to the sulfur-bearing layers at depths of 150–750 m and escapes through holes in the lower end, melting the sulfur. Compressed air in the inner pipe pushes molten sulfur up through the middle pipe. On the surface, sulfur is stored in solid form in huge vats, but in recent years delivery to the market has increasingly taken place in specially designed heated tanks, carried by truck, railroad, barge or ocean freighter, in which the sulfur remains liquid.

Italy, once a major world producer of sulfur, primarily in the Caltanisetta district of Sicily, where native sulfur occurred in sediments associated with gypsum, now has an insignificant production from sulfur ores. Most of its sulfur is obtained from pyrites. The only important producers of sulfur from sulfur ores are the Soviet Union and Poland.

The increasing emissions of sulfur dioxide from smelters, thermoelectric power stations, and chemical plants have become a serious environmental problem in the industrial countries and governments have been forced to enact restrictive legislation. Since governments are interested in the competitive position of their manufacturing industries in the international market, legislation has been combined with favorable terms for investments in sulfur recovery equipment, subsidies for the transport of sulfur to the market and so on. For that reason by-product sulfur is likely to account for increasing shares of the total production. Some Frasch plants may have to be closed.

Canada accounts for almost half the sulfur recovered from the roasting and smelting of metallic sulfides (copper, zinc and lead), and the refining of 'sour' petroleum, which makes it the second largest sulfur producer in the world and a leading exporter. The rapid expansion of sulfur production is primarily a result of increased demand for natural gas. The rated capacity of Canadian sulfur recovery plants for treatment of sour natural gas in 1970 was 5.6 million tons or more than the recovered sulfur (4.4 million tons). Unit trains operate between the recovered sulfur plants in Alberta and a common stockpile in Vancouver, from which 24 participating sulfur producers draw for export sales.

The Athabasca tar sands of northeastern Alberta are one of the world's largest potential reserves of crude oil and of sulfur (4.5%). One of two pilot plants in the area has a capacity of 2.5 million tons of synthetic crude oil and 120 000 tons of sulfur. France and the United States are the other two large producers of recovered sulfur. The French output comes from natural gas plants.

About 1970 Iran was building large such plants at Bandar Shahpur and Karg Island. The former is part of a petrochemical complex. Iranian reserves of sulfur in its natural gas deposits are very large.

Large producers of sulfur from pyrites are the Soviet Union, Japan, Spain and China. It is also important in Italy, Cyprus and Norway.

Fertilizers

The fertilizer minerals occur naturally in the soil but are removed systematically with the crops. The high-yielding crops of modern agriculture presuppose a retained mineral balance in the soil. Farmers and agricultural researchers must have their attention fixed not only on the hydrological cycle, but on a nitrogen cycle, a phosphorus cycle, a potassium cycle, and a calcium cycle as well. They are integrated parts of a general system that dictates the conditions for all living things on the Earth (the biosphere).

The natural fertility of a soil may be great, but sooner or later the soil will be exhausted if minerals are not supplied through manure, compost, green manure or commercial fertilizer. In the American Middle West crops were taken year after year from the pioneer days to the mid-1930s with practically no fertilizing. Only at the end of the hundred year long period did yields decrease somewhat. On this exceedingly fertile land yields have doubled or trebled in the forty years before the mid-1970s thanks to modern fertilizers.

In China and Japan farmers succeeded in keeping the yields at a high level for millenia with manure and compost. Until recently, 'collection' took place even in the metropolitan households. The 'night soil' was distributed to farmers around the cities. Early travellers in China could report that cities ahead could be sensed long before they came over the horizon by higher crops and greater intensity of the green. The introduction of chemical fertilizers in Japan, the first country in East Asia to be modernized, meant that already high yields were doubled between 1880 and 1960. A similar stepping up in yields was started in China in the 1950s.

In India, where a large human population has competition from a tremendous population of cattle for the limited arable land and where cow dung because of forest depletion is being used as a fuel, soils have been exhausted and yields extremely low. (Yields have been extremely low in comparison with those of mid-latitude countries but not necessarily with those of other tropical countries.) After the introduction of the Green Revolution in the second half of the 1960s the fast growing fertilizer consumption increased at a record level. Fertilizer consumption (production and imports) and family planning have had the highest priority in the latest Indian five-year plans. The promising development was hit hard by the oil crisis of 1973/74. But the counter-measures taken by the Indian government (increased exploration for oil and gas) may in the long run prove beneficial for the national economy.

It can be shown that the marginal increase in yield per ton of fertilizer is much higher in India than in countries with a high fertilizer consumption. A rapid temporary switch of marginal fertilizer quantities in favor of India, e. g. such used on lawns, cemeteries, and golf courses in the rich countries, would make large aid shipments of grains a year later superfluous. However, a world government making rational decisions about the world economy does not exist. No one can direct fertilizers to the areas where they would add most to the world crop. In theory, the market should function as such a global director. In practice, however, the 'fixers' who could give the market forces a humanitarian profile do not exist but rigid, prestige-seeking bureaucracies in several places (the industrial countries, but primarily the United Nations and the receiving countries). Agriculture is the sector in which the world is furthest away from free trade.

Since the food supply in the tropical world with its rapidly growing population will remain in the focus of the political interest at least during the remaining decades of this century the supply of fertilizers will also play a prominent part.

Nitrogen (N), phosphorus (P), and potassium (K) are the most important elements added to the soil in the form of commercial fertilizers to improve the growth of plants. They are sold separately or in combinations. Other macronutrients, needed in quantities between some kilograms and some quintals per hectare are calcium, magnesium and sulfur. Some nutrients are needed only in quantities expressed in grams per hectare (micro-nutrients, trace elements), e. g. manganese, copper, zinc, boron, iron, molybdenum, and cobalt. If they are missing in the soil they may be added separately or mixed into standard fertilizers.

Commercial fertilizers were first used in Europe in the 1840s. Justus Liebig of Germany in his pioneer work drew attention to the need to analyze each species and each soil type to ascertain the right fertilizer mix. Fertilizers fitted well into the land-hungry and labor-intensive agricultural system of northwestern Europe at the time, just as farm-machinery fitted into the contemporary farming system of the American Middle West. A hundred years later the European farmer started to use machinery on a large scale and the farmer of the Middle West applied fertilizers. Urbanization and international migration in the meantime had achieved an economic and technical convergence between the two regions.

Both in Europe and the United States, with their capital-intensive but labor-extensive agriculture, the plateau should be rather close where additional inputs of fertilizer does not result in an additional yield increase. For the entrepreneur (the farmer) the interesting point is where an additional fertilizer dollar yields just one additional crop dollar. To go beyond that point does not make economic sense although the yield may still increase in physical terms. The fertilizer prices (which primarily are influenced by the price for energy) are crucial in that equation.

In the tropical countries, which are in the very beginning of their transition to a capital-intensive agriculture, the marginal yield-increase per ton fertilizer is much higher than in the industrial countries. One more ton of fertilizer yielded three more tons of grain in Western Europe and the United States in the mid-1970s but no less than six tons in India.

Since fertilizers have a low value per ton, transport costs account for a large share of the price. The assembly costs of energy and raw materials and the distribution costs of the products are of great importance for the location of the production units.

Nitrogen

Nitrogen is used in explosives, plastics, resins, synthetic fibers, feed, pulp and paper and so on, but $3/4$ of the peace-time production goes to commercial fertilizers. It is the most expensive of the macro-fertilizers and the one that is used in largest quantities. To obtain nitrogen from the air is an expensive process but it is cheaper than mining natural sodium nitrate (Chile saltpeter) in the desert of northern Chile, or obtaining nitrogen compounds as by-products of coke ovens and gas works or as organic materials, e. g. dried fish refuse, dried blood or other waste from slaughterhouses.

Nitrogen is an essential component in all plant and animal tissues and plays a leading part in the development and functions of protoplasm. It is especially important in tissues concerned with growth and reproduction. Plants use nitrogen to build amino acids which are the building blocks of proteins and other organic compounds that later are used by animals and microorganisms. One such organic compound is chlorophyll which is essential to the photosynthesis of plants. Nitrogen fertilizer is directly reflected in a characteristic shade of the green.

The fusion of nitrogen and other elements, fixation, requires much energy. An increase in the price of energy is directly and strongly reflected in the price of nitrogen fertilizer.

Nitrogen that plants obtain from the soil may have several origins: the disintegration of organic material, fixation of molecular nitrogen through microorganisms that live independently or in symbiosis, commercial fertilizer and so on. The unique characteristic of the bacterium rhizobium, which lives in symbiosis with plants of the pea family, to which it contributes nitrogen that it takes from the air, has long been utilized in agriculture, where pea plants are regularly included in the crop rotation. Intense research is carried on to adapt

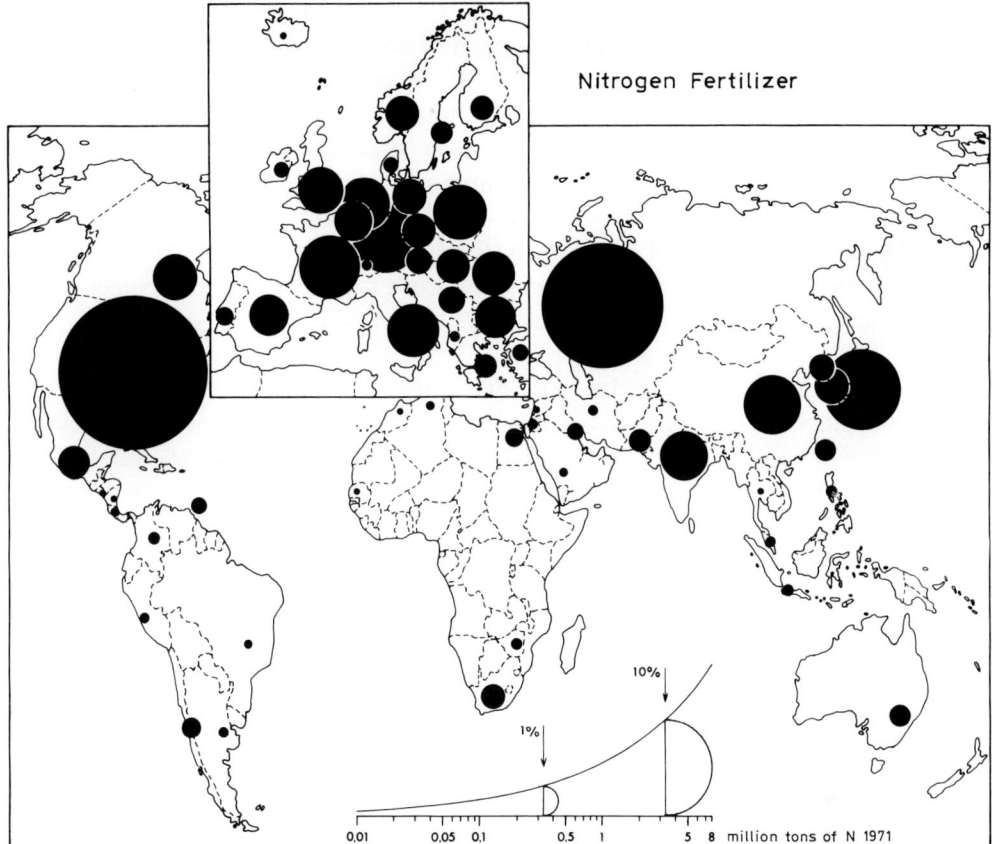

Fig. 40 – Fertilizer production is one of the fastest growing mineral industries, not the least in tropical and subtropical Asia, with a global doubling time of less than ten years. Europe no longer dominates world production as in the interwar years.

rhizobium to a life in symbiosis with the most important grains. This form of plant breeding would, if successful, save much energy.

Chile saltpeter – Until the First World War, Chile was the most important producer of unorganic nitrates. At the turn of the century, Chile had $^2/_3$ of the world trade in nitrogen fertilizer. The nitrate deposits of the Atacama Desert are irregularly spread over a zone, 720 km by 15–80 km, in the slightly sloping pampa between the Coastal range and the Cordillera at an altitude of 1 200–2 200 m. In the beginning, mining methods were simple. A caliche was blasted and the lumps were loaded unto railroad cars for transport to the oficina. Here they were crushed and dumped into boiling water after which the dissolved sodium nitrate was precipitated in cooling vans. Iodine was obtained from the solution and this by-product normally accounted for 70% of world production.

The nitrate was sent in sacks to the ports. For many decades the nitrate flows between Chile and Europe – North America (the east coast) were among the leading trades in world shipping. Peaks in production were reached 1916–18 and 1929 with some 3 million tons. Between 1880 and 1930 Chile levied a tax of US $12 per ton on its nitrate export. Competition from other sources of nitrate forced a heavy mechanization of mining and transportation. A new process (the Guggenheim process) allowed mining caliche of lower grade and recovery of a higher percentage of nitrate. The end product was small pellets that were handled in bulk instead of in sacks. But Chile supplies a declining percentage of the world's nitrate, now less than 0.3%.

Norwegian saltpeter – In 1905, two Norwegians, K. Birkeland and S. Eyde, developed the arc process of extracting nitrogen from the atmosphere. Factories were built at Rjukan and Notodden, adjacent to remote water power stations. Norwegian saltpeter (calcium nitrate) became the first synthetic competitor of the natural Chile saltpeter.

Ammonia – The German Fritz Haber in 1909 made an important contribution to industrial chemistry with his high pressure synthesis for making ammonia from nitrogen and hydrogen. He received the Nobel prize in 1918 for his invention. During the First World War, Carl Bosch developed the process for industrial application. In 1917 the Haber-Bosch-method accounted for 45% of Germany's production of nitrogen compounds. At armistice, Germany was independent of imports for its ammunition factories. The Haber-Bosch-method made possible the long German participation in the war.

The Haber-Bosch-method could also be used in fertilizer factories. It was cheaper than the Norwegian method and in 1928 Norsk Hydro also switched to Haber-Bosch. Ammonia can be used directly as a fertilizer but as a rule it is a raw material for making other nitrogen compounds: ammonium nitrate, ammonium sulfate, or ammonium phosphate, nitric acid, and urea (a compound occuring in urine, also called carbamide). To make one ton of ammonia requires 1500 kWh of electricity. The energy price is important for the location of ammonia factories. Nitrogen is a ubiquity, obtained from the air. Some carbohydrate provides the hydrogen. Earlier coke oven gas was used but now the petroleum fraction naphta and, above all, natural gas are much cheaper. Oil and gas may provide both hydrogen for ammonia and fuel for the thermoelectric power plant.

World production of nitrogen contained in fertilizer compounds in 1970 was 30 million tons after a very rapid growth in the 1960s (12% a year). Expansion was most spectacular in the Soviet Union, Asia and North America while Europe continued to grow at a fast clip in spite of a high point of departure. Africa had a slow growth and South America a decline.

African experiments with fertilizers have given good results in some countries, e. g. Nigeria. However, low product prices in the agricultural districts and high fertilizer prices as a result of a poor road net provide low profits to the farmers. But inflation in food prices on the world market after 1972 will probably lead to more emphasis on having domestic farmers provide import substitution for food normally imported to the coastal cities. Several countries in Africa have an excellent raw material base for a domestic fertilizer industry, not the least oil and gas.

All central plan economies of the Eurasiatic continent expanded at an above average rate from a low point of departure. Countries involved in the 'Green Revolution' (Mexico, India, Pakistan, Indonesia, the Philippines and others) were in the same situation. Consumption of nitrogen fertilizers grew even more drastically than production in some of these countries, e. g. Turkey, India, and Indonesia, where large imports made up for the difference. India in 1970 had less than half the production of West Germany but a larger consumption.

In the United States fertilizer consumption was strongly concentrated to the Atlantic coastal states until the early 1930s according to the principle: heavy rainfall, more leaching, more fertilizer. Tobacco, vegetables and other regional crops reacted favorably to fertilizer inputs. Acreage restrictions, a feature of the New Deal in American agriculture, led to application of fertilizers on a large scale also in the Middle West. Fertilizers are the main factor behind the doubling or trebling of yields between the mid-1930s and the 1970s.

Because of the central role of natural gas in modern ammonia production, the petroleum exporting regions (Persian Gulf, Venezuela, North Africa) are rapidly expanding their ammonia producing capacity.

Phosphorus

Phosphorus has many uses, e. g. in the making of chemicals to be used in detergents, medicine and food, but ¾ of the consumption is accounted for by agriculture which uses phosphorus pentoxide (P_2O_5) as a fertilizer. Phosphorus is part of all cells. Among plants the concentration is strongest in the seeds and among animals in the skeleton. Good access to phosphorus speeds up the ripening. Deficiency in phosphorus will occur after a few years of cultivation, especially with grains and cattle in the system. Already in the 17th century, bone meal was used in England to add phosphorus to the soil, although at the time no one knew why. The Indians of North America put fish on their corn fields, which added both nitrogen and phosphorus to the soil. The most important sources are phosphate and apatite rocks, which are produced in larger quantities than other mineral fertilizer materials except limestone for lime. They supply more than

80% of the world's phosphate fertilizer. The Thomas and the basic open-hearth steel furnaces utilizing pig iron made from high phosphorus ore supply less than 15% and organic material (guano, bones and other refuse from slaughter houses) less than 5%.

Apatite occurs in igneous rocks, e. g. in the Fennoscandian pre-Cambrian shield. It is mined on a large scale at Kirovsk in the Kola Peninsula, the world's largest apatite field. Some mining also occurs in North Vietnam. The apatite deposits associated with the iron ore fields in Lapland were mined when Sweden was cut off from overseas supplies during World War II but are normally not competitive. Apatite supplies less than 15% of the world's mined phosphate.

Sedimentary deposits dominate world production of phosphate. They are thought to have two types of organic origin, either accumulations of bones and excrements from land animals in caves and dry areas or from marine animals on the sea bed. Raw phosphate usually occurs as pebbles embedded in calcium carbonate. Sometimes the binder has weathered away, sometimes the mass is petrified. Very large reserves of phosphate are found at shallow depths on the shelves off the coasts of many countries in the world.

United States – The production of raw phosphate is dominated by the United States with over $^2/_5$ of the world's total. The industry dates back to 1867 when low-grade deposits in South Carolina were exploited. The high-grade Florida deposits, which now supply 90% of the American output, have been the leading American field since 1887. Workable beds are located in a 180 km long and 50 km wide belt in the west-central part of the state with the chief producing area near Tampa. These 'land-pebble' deposits are up to 13 m in thickness, unconsolidated and ranging in fragment size between silt and pebble. They are easy to work with modern dragline shovels that scoop up a carload of phosphate at a time. North of Tampa are hard-rock phosphates with a higher phosphate content but more expensive to work.

Tennessee and some western states have some production, in Tennessee from open pits and in the west primarily from shaft mines.

Tab. 52 – World Production of Raw Phosphate 1972
(million tons)

United States	38.5	Tunisia	3.4	Togo	1.9
USSR	22	South Africa	2.0	Senegal	1.3
Morocco	15.1	Nauru	1.9	Christmas Isl	1.1
World	94.4				

Source: SY1973.
World Export 1973: Morocco 16.5, United States 12, Soviet Union 6 (IHT 10–11/11 1973).

North Africa leads the world in known phosphate reserves but since large parts of the world are still incompletely investigated the current reserve situation may be of limited interest. The recent large finds in Spanish Sahara are a case in point; they focused the world's interest on this 'forgotten" territory.

The rich deposits of Morocco have been known since 1912. A 140 km railroad between Khouribga and Casablanca was opened in 1921. Another production area was opened in the 1930s at Youssoufia. It has an 80 km railroad to Safi. A third mining area went into production in the early 1970s at Ben Guerir, 160 km east of Safi. The state-owned Office Chérifien de Phosphates (OCP), because of its monopoly position in the world market, has succeeded in raising its prices as much as the oil monopoly OPEC. Prices went up from US $14 a ton in November 1973 to $63 a ton in July 1974.[26] The OCP is Morocco's leading industry and accounts for 20% of her BNP. Morocco expects to export 50 million tons of phosphate a year by the end of the century. Two new phosphate ports are under construction at El-Jadida and Essouira, the latter with credits from the Soviet Union. Morocco is expected to export 10 million tons of phosphate a year by 1990 to Soviet ports on the Black Sea. This phosphate will be cheaper for the Soviet Union than exploiting deposits in the Kola Peninsula or Central Asia. The Soviet Union is expected to receive annually 10 million tons of phosphate during a 25-year period. As payment, the Russians will build a phosphate terminal, a dam, a new phosphate mine and deliver lumber and petroleum.

Since the United States and the Soviet Union use more and more of their phosphate domestically, Morocco's monopoly position in the world market is strengthened. In 1974, Morocco shipped 21 million tons and received US $1 000 million in revenue; two years earlier the corresponding figures had been 15 million tons and US $152 million.

The time sequence of the start of raw phosphate production in the Maghreb states reflects the proceeding of French colonization. The earliest find was made in the oldest colony, Algeria, where phosphate was mined already in 1873. The oldest mines have been closed because of low phosphate content but in 1966 mining was started at Djebel Onk where the phosphate containing layer is 30–35 m. It is thought to be one of the largest deposits in the world.

In southern Tunisia raw phosphate was found in 1885. Before the Second World War, Tunisia was the leading producer in North Africa but is now far behind Morocco. The largest mines are found in the Gafsa area some 200 km from the shipping terminal at Sfax. A smaller mine is south of Tunis with terminal at La Goulette. At Sfax and Gabes (and at Safi, Morocco) are export-oriented superphosphate factories.

Spanish Sahara with only 75 000 inhabitants, most of them nomads, has very large phosphate reserves at Bou Kraa, surveyed after 1963. The forgotten corner of Africa, became a Spanish colony by default. No other country was interested. The prospects of finding large mineral resources in the wayless desert area made the neighboring states, especially Morocco, develop lively diplomatic activity to get the area incorporated into the neighbor countries when it gained independence from Spain in 1976. Morocco even staged a march into Spanish Sahara at the end of 1975 and reached an agreement with Spain about taking over the administration of most of the territory. However, this was contested by Algeria.

Among lower rank phosphate producers in Africa, now more important than Algeria, are South Africa, Togo and Senegal. Egypt is an important producer and consumer of both nitrogen and phosphate fertilizers.

Soviet Union – In addition to its apatite deposits, already mentioned, the Soviet Union has raw phosphate, primarily in the large mine at Karatan near Dzhambul in southern Kazakhstan but also at several places in the European part of the country.

Asia – The most important producing areas are in the southwest straddling the border between Jordania and Israel, in the Ammam district and in the Negev desert. Israel exports phosphate from Oron and Arad by way of Ashdod on the Mediterranean and Eilat on the Gulf of Aqaba. The Jordanien exports from Ruseifa and El Hasa move primarily by way of Aqaba. The phosphate deposits of Iraq, Iran and Saudi Arabia – like other mineral deposits in this part of the world except petroleum – are incompletely known. Christmas Island in the Indian Ocean and China are among the medium-sized phosphate producers.

Oceania – In the Pacific, Nauru and Ocean Island are the leading producers with Australia as the most important market. In the world's driest continent, superphosphate has long been by far the leading fertilizer. The economy of the two phosphate islands is dominated by the phosphate mines, which originally were worked by contract workers brought in from China. During the Second World War the installations were almost completely destroyed.

Latin America – Peru with its guano deposits has a tradition of fertilizing and of fertilizer export. Today guano is primarily used in the highlands, where it is sold without charge for transport costs, while primarily commercial fertilizers are used in the coastal plains. Guano production decreased rapidly in the 1960s as a result of overexploitation and the death of birds as a result of an expanded fishing industry. Peru has large raw phosphate deposits near Talara in the Sechura Desert.

Potash

Potash or pottasium oxide (K_2O) is used in the manufacture of soap, explosives, matches, drugs, glass, paper, paints, photographic material and in many other products but $9/10$ is used in fertilizers. In comparison with nitrogen and phosphorus, potassium is common in the earth's crust. Like the other two elements, potassium is essential to all living things. If a field produces crops year after year, yields will eventually decline if potassium is not added.

Exploitable potassium is almost exclusively obtained from brine and salt deposits containing potassium minerals (carnallite, sylvite, and kainite). Like related minerals, common salt, halite, gypsum, and anhydrite, they have been precipitated from salt solutions in ancient seas; such deposits are known as marine evaporites. Some 95% are mined at sizable depths from vast deposits in Germany, the Soviet Union, Canada, United States, France, and Spain.

Other potassium sources that formerly were widely used and which still have non-commercial importance in large parts of the world are vegetable matter like ashes, sugar beet tops, kelp, and sun flower stalks as well as animal matter like waste from the washing of wool.

Europe — The vast potassium deposits around Harz in Germany have been an important raw material base since they were first exploited in 1861. The potassium minerals are here found in layers of 60–90 m at a depth of 300–750 m and are worked in an extensive system of shafts and tunnels. Best known are Stassfurt and Mansfeld northwest of Halle in Germany (E). Brown coal is found near potassium and other salts which has changed the region to a center of the chemical industry (the Leuna works at Merseburg south of Halle). The German potassium deposits are in the middle of a vast fertilizer market with land and water routes to ports on the North Sea and the Baltic. They are surpassed only by the vast deposits of the USSR and Canada and account for nearly 40% of the world production.

The French deposits at Mulhouse in eastern Alsace cover an area of 180 km² and are thinner than those in Germany. They are located at a greater depth, 480–840 m. Like the German mines they are served by an excellent canal, river, road, and railroad net. Until 1918 Germany had a potassium monopoly since the deposits of Elsass were German between 1871 and the end of WW I.

The Spanish deposits northwest of Barcelona are of high quality but the layers are folded. Reserves are large but production limited. Israel's rapidly increasing production is obtained from the evaporation of the water of the Dead Sea, evaporated for potash as well as for other salts. Both sun and thermic evaporation have been tested. (Sun evaporation on a large scale is also being used for a new project on the Great Salt Lake in Utah.)

Deep-lying potash deposits were found in 1963 near Gdansk in Poland. Like the German deposits, they belong to the Permian (Zechstein). Production in this deposit that extends under the Baltic was started in 1973.

North America — Before the First World War, the United States was dependent on Germany for its potassium supply. During the war, prices increased tenfold and the government encouraged production from salt lakes in Nebraska and California, but these installations were discontinued after the war. From the end of the 1920s, German and French producers established monopoly pricing which helped the development of the industry in the United States. In the 1930s, oil drillers found the large Carlsbad deposit in New Mexico at a depth of less than 300 m. At greater depth, this deposit extends over a very wide area. Carlsbad accounts for 85% of the American production. The rest comes from salt lakes in Utah and California. Large underground deposits are also found in Utah and Michigan.

Tab. 53 — World Production of Potash (K_2O) 1972 (thousand tons)

USSR	5 498	Germany (E)	2 458	Israel	566
Canada	3 747	United States	2 412	Spain	490
Germany (W)	2 845	France	1 760	Zaire	474
World	20 490				

Source: SY 1973.

The vast potash reserves found by oil drillers in Saskatchewan, Canada, in 1943 are Devonian and thus the oldest being exploited. The potash layers stretch from northwestern North Dakota through southern Saskatchewan. Most of the strategically located Canadian production is shipped across the national border to Illinois, Iowa, Indiana, and Minnesota, the heaviest potash consuming states in America. The United States, before 1960 a net exporter of potash, in 1970 had gross imports almost equalling the shrinking domestic production.

USSR — The Soviet production of potash increases rapidly after a long period of notoriously poor utilization of fertilizers. The old mines at Solikamsk in the Perm Basin of the Urals, for which the Permian period was named, are the most important deposits but they are getting a rival in the new mines at Soligorsk in Belorussia.

Calcium

Calcium (Ca) is a macro-nutrient that crops use in relatively moderate amounts. But the calcium base calcium carbonate, $CaCO_3$, has another important function. The roots of plants breathe and form carbon dioxide which, with ground water, forms carbonic acid. Also sulfuric acid and nitric acid are formed. Soil is neutral or faintly alkaline as long as it contains lime, but when all lime particles have been washed out it turns acid. Supply of lime to the acid soil (low pH-value)[27] neutralizes the acids and increase the value to the neutral 7, which is most advantageous for most plants.

From a geographic point of view, lime production is of relatively small interest since limestone is a near-ubiquity. Agriculture normally gets its supply of lime through relatively short transports. Lime is also used as a flux material in the iron and steel industry. The Great Lakes and the Baltic have rather large traffic flows of limestone.

Abrasives

Abrasives include natural and synthetic materials used for grinding and polishing.

Industrial diamonds

Industrial diamonds are used because of their extreme hardness for precision grinding and cutting, for bushing in fine wire-drawing machines, and for optical and precision instruments. Diamond is pure carbon (C) formed under extreme pressure and high temperature. Industrial diamonds are not suitable for jewels which catch a higher price. They may have flaws, unsuitable color or be of small size. Bort or small fragments, produced at diamond cutting centers, are a special type. Since 1956 synthetic diamonds have been made. Other synthetic abrasives of similar hardness have been developed. One is Borazon, the trade name of boron nitride, first produced in 1957. It is more resistant to high temperatures than diamond.

The mother lode of diamonds is kimberlite, a green-black, ultra-basic igneous rock in which diamond is crystallized under high temperature and pressure, presumably at great depth. The kimberlite forms cylindric intrusions, probably volcanic necks, known as 'pipes'. All pipes do not contain diamonds, but kimberlite is the only mineral in which they have been found. Secondary deposits have always been important: river beds, shores, and glacial gravel. Such diamond layers vary in age from recent to pre-Cambrian. The older ones have often been consolidated into conglomerates. Often the position of the original pipe remains unknown.

World production of natural diamonds in 1970 was 42 million carat of which 28 million industrial diamonds and 14 million jewels. Output has increased substantially since the large African mines were opened in the 1860s. The consumption of jewels increased by 5.4% a year in the fifteen-year period before 1970 and of industrial diamonds even faster. Expressed in carat, the consumption of industrial diamonds surpassed that of jewels for the first time in 1929. By value, jewels are still as important as industrial diamonds. The only large producer outside Africa in 1970 was the Soviet Union with more than 20% of the industrial diamonds and more than 10% of the jewels. The United States produced 13 million carat of synthetic diamonds in 1970; other producers were the Soviet Union, Japan, Sweden, Ireland, and South Africa.

South Africa and Namibia still produce almost half the jewels of the world but only 15% of the industrial diamonds. The Kimberley district, where production started in the 1860s, is still most important. The Finsch mine, 160 km west of Kimberley, went into production by 1970, the first new pipe in South Africa since the Premier mine near Pretoria was opened in 1903. It is the only one in the country outside the Kimberley district. Marine diamonds are obtained off the coast north and south of Port Nolloth near the border of Namibia. A rich diamond pipe at Orapa in Botswana has been mined since 1971. The Oppenheimer company De Beers Consolidated Mines, with headquarters in Kimberley, ist the leading diamond group with an annual production of about 11 million carat.

Zaire dominates the mining of industrial diamonds with over 40% of world production. The largest production occurs at Bakwanga and other places of the Kassai Province, where diamonds are found in alluvial deposits. The diamond-carrying gravel is dredged and carried to a washing plant. Other important producers in Africa are Ghana, Angola, and Sierra Leone.

The Soviet Union was an importer of diamonds until the deposits on the Lena River is eastern Siberia, discovered in the 1950s, were mined. The city of Mirnyj was founded to serve the diamond district of Yakutia. The Mir-pipe is kimberlite like the prototype at Kimberley. Alluvial diamonds have long been mined in the Vishera district of the Urals.

Brazil was the dominant world producer before the large discoveries in Africa. It now accounts for less than 1%. Most carbonados (black diamonds) and ballas are mined in Brazil (Bahia and Minas Gerais).

Aluminum Oxide and Other Natural Abrasives

Corundum, natural aluminum oxide, is second only to diamond in hardness. As emery it has special uses in industry, e. g. for grinding optical lenses. For emery, opaque dark corundum is used. Pure, translucent forms are valued as jewelry: sapphire (blue) and ruby (red).

The United States bought all her corundum from Rhodesia before the economic sanctions by the United Nations against that country were enforced 1968. In 1970, the Soviet Union accounted for 80% of known production, followed by India and South Africa, but Rhodesia may have continued her production on a high level. Official statistics provide no answer to that question.

Emery is a mixture of corundum and iron oxide which is a little less hard. It is also used as an antiskid on floors, stairs and sidewalks. The Greek island of Naxos is known for its emery. The United States has a deposit at Peekskill in New York. Other producers are Turkey and the Soviet Union.

Garnet is a natural silicate abrasive, used for grinding cloth, lapping of metals, and for grinding and polishing of optic lenses and plate glass. It is found in New York and Idaho. Garnet is also used as a precious stone.

Corundum, emery and garnet are rapidly being displaced by synthetic materials: molten aluminum oxide, silicon carbide, boron carbide, tungsten carbide, tantalum carbide and titanium carbide. Synthetic abrasives are produced in much larger quantities than the natural ones.

Many other materials are used for industrial grinding. Large quantities of sand, sandstone and quartz are mined for use in sandblasting. A finegrained, siliceous stone mined near Hot Springs, Arkansas, is still used as oilstone, which is in demand for sharpening fine instruments. **Tripoli,** a soft siliceous material from Arkansas and Oklahoma is used as an abrasive and filler. The same holds for a similar amorphous or soft silica from southern Illinois and **rottenstone** from Pennsylvania. Pumice, diatomite and various clays are used in fine grinding materials and hand soap.

Boron

Boron (B) occurs naturally in borax and boric acid. It is obtained from dry lakes in desert areas with their own drainage basin. Pure boron has no practical use but its compounds are important. About $1/3$ is used in glass, e. g. in heat resistant glass and fiber glass. Some 15% are used in enamel for coating on kitchen sinks, kitchen stoves, refrigerators, freezers and so on and a similar quantity in soaps and detergents, including toothpaste and mouthwash. Agricultural chemicals also take 15%, partly as an addition to commercial fertilizers to supply the soil with boron, which is an essential element for the plants and partly as an ingredient in herbicides. Boron is also used in steel alloys to increase hardness. Some boron compounds, such as boron carbide (used as a moderator in nuclear reactors, as an abrasive, and as a refractory), titanium boride and tungsten boride are among the hardest substances known.

California dominates world production of boron minerals. Most is obtained from opencasts in dry desert lakes in Inyo and Kern Counties and some from Searles Lake in San Bernardino County. In the 1970s, a mine was opened in Death Valley. Almost half the American output of boric acid and borax, some 500 000 tons in 1970, is exported, primarily to the Netherlands and Japan. Europe is supplied from a central warehouse in Rotterdam. The most important ocean terminal is at Wilmington in the Los Angeles area.

Turkey is the second-largest boron producer. The Soviet produces boron from deposits north of the Caspian Sea. Argentina (Salta) and China (Sinkiang) also obtain boron.

Mineral Raw materials in International Trade

Developed countries (DCs) are wealthy because they exploit less developed countries (LDCs), buying their raw materials, including energy and minerals, at bargain prices. LDCs are less developed because the prices of their raw materials are kept at an exorbitantly low level by the DCs. This Marxist dogma in recent years has gained credence in wide circles. In several DCs, among them Sweden, massmedia have a Marxist bias that in no way reflects the Marxist position in the electorate. Massmedia people with a Marxist jargon have been conspicuous in media with a strong impact on public opinion, primarily TV and newspapers with a large circulation.

The exploitation credo is not limited to Sweden and other DCs as testified by any number of political statements by leaders of LDCs, some of which in 1975 might better be termed newly-rich nations. These politicians often had their education at European or American universities in close contact with the radical debate of the day.

Is the exploitation credo true or false? The answer should be found in the border area between economic and political geography. In market economies almost all international business transactions are made by business firms, as a rule privately owned but sometimes with a controlling goverment interest. However, firms with a large or controlling government ownership (e. g. the British oil company BP or the French automobile-manufacturing group Renault) do not act differently from those which are completely in the private domain. Business firms vary considerably measured by the role international transactions play in their total turnover. Some are pure domestic firms, others import rawmaterials, parts and components and sell some of their output abroad and a third group not only trade with partners abroad but also manufacture outside their home country. They are the multinational corporations (MCs) for whom exports and imports may be transactions between subsidiaries of the same group.

The business firm is the agent in economic geography. Survival and longterm growth are the two major goals of the business firm. Exploitation of foreigners cannot be a goal per se. Such exploitation must take the form of underpayment of foreign employees, charging customers too much or paying suppliers too little. It seems legitimate to assume that subsidiaries of MCs in accordance with their two major goals follow the rules laid down by the host governments. MCs will have to pay at least market wages and salaries and charge or pay market prices.

Since many MCs in small LDCs are monopolies, it should be reasonable to assume that they may charge too much for their products or, which is our primary concern in this book, pay too little for their raw materials. But the blame should then be put on the host government for having too lax anti-trust legislation, not on the company for following the rules of monopoly pricing. Political geography does not enter this example: it would be difficult to find an instance in which a home country of a MC has tried to prevent a host country of a subsidiary from enforcing anti-trust laws.

Raw materials and political geography. Political geography or geopolitics deals with the geography of power[28]: in focus is the state and its relations to other states seen from a geographical point of view. Modern textbooks in political geography agree about the core of the subject. But in an IGU Bulletin (1974:2) the International Geographical Union states that political geography lacks focus.

The statement of the IGU seems strange at a time when one of the reasons for the 1973 oil crisis and all its disrupting consequences was the lack of geopolitical analysis in the oil-importing as well as the oil-exporting countries. Events after October 1973 indicate that all countries had too few practitioners of the geopolitical art.[29]

It is an important task of the government in a modern state to guarantee a steady and uninterrupted flow of energy to the various sectors of its economy. Market economy countries find it natural to let private, cooperative and state groups in competition organize this flow. But it is an intrinsic right of the state to formulate the rules of the game, to develop an energy policy just as of old it has had an agricultural policy. Energy corporations as well as energy consumers look to the state for an enunciated policy for the supply of energy. The lack of a stated policy must be interpreted as a laissez-faire-policy. Such a course of inaction was typical for most industrial countries during the twenty-year-period before 1973. Deviations were rather small: the American import quotas for oil to protect domestic production of energy and socially motivated discrimination of oil imports into several countries in Western Europe to protect the coal miners from too rapid decline in the number of jobs were the best known examples. The laissez-faire-policy led to low and decreasing energy prices, not the least in extreme free-trade countries in the petroleum sector like Japan, Italy and Sweden. It also led to vast profits in countries with low production costs (high land rent), primarily several countries of the Persian Gulf-region in which the geological structures are exceedingly favorable and the average oil well produces several hundred times as much as the average American well. Towards the end of the period Japan and Western Europe for the first time in history had lower energy prices than United States. The laissez-faire-policy led to extreme vulnerability as demonstrated when the cartel of the exporting countries applied reckless monopoly pricing.

The paragraphs above should not be seen as a plea for a nationalistic economic policy. The state should guarantee an uninterrupted flow of energy even in times of crisis but it should not necessarily encourage energy production within its own borders. Oil-importing countries in regional cooperation with other oil-importing states should ensure that they are not too dependent on one oil-exporting region or one energy-exporting cartel. The laissez-faire-policy may lead to low energy prices in the short-run but at the risk of disruptions and high prices in the longer perspective.

The economic and geographical deliberations of the integrated production and distribution groups should lead to a concentration of oil production to the regions with the lowest production costs. This was what happened in the twenty-year period before October 1973. The Organization of Petroleum Exporting Countries (OPEC), formed in 1960, had not yet started to function as a multi-state monopoly of the leading petroleum exporting countries. It was still a typical pressure group.[30] The integrated multinational oil corporations wanted to produce as much as possible in the low-cost areas and as little as possible in the high-cost areas.

History has proved oil men to be far from the omniscient people assumed by economic theory: instead of starting oil production in the best area and from there proceeding to less and less favorable districts they have done the opposite which is true also of other mineral producers. The oil producers started in the high-cost Ohio River Valley of the United States (1859) and not in extremely low-cost Saudi Arabia (1932). The former is the home area of leading multinational oil corporations (the Standard Oil groups and Gulf), the latter is a typical host area of multinational corporations but no such groups were born there.

It is reasonable to give consumption primacy over production. Production occurs to satisfy consumption needs and not the reverse. Likewise the state has primacy over the business firm. The state can take over the firm but not the opposite. It is logical to assume that the geopolitical deliberations of the importing countries are the key to future oil production patterns and oil trade flows. Any disturbance in the delicate balance of a laissez-faire-situation should lead to a search for imports from alternative regions.

Recommendations for cooperation among the oil-importing countries had not been lacking before October 1973. Writers on the economics and geography of oil, like Peter Odell, M. A. Adelman and Walter J. Levy had come out in favor of such cooperation. The initiative was taken by the United States after the Arab oil embargo was imposed in 1973. The International Energy Agency (IEA) was conceived in February 1974 and a year later had 18 member countries: Anglo-America, EEC except France, Japan, Australia, New Zealand, Spain, Sweden, Switzerland and Turkey. Its staff is attached to the headquarters of the OECD in Paris. The agency has established an oil-sharing plan. Members have pledged a 10-percent reduction in oil imports. In March, 1975, IEA agreed on setting an oil floor-price which should be a sine qua non for oil-importing countries that have faced an embargo on oil exports and drastic monopoly pricing by a multistate monopoly. A guaranteed price is a necessity for all decision makers facing investment decisions in alternative petroleum producing areas as well in alternative energy forms.

Petroleum and other minerals. The oil crisis starting in October 1973 probably marks the end of laissez-faire policy in the international petroleum trade. The global marginal cost for petroleum production was never allowed to play its full role but the continuous decline in real petroleum prices in the twenty years before 1970 indicates a strong trend in that direction. New low-cost producing wells in new countries were opened faster than required by the increase in demand. High-cost wells in old oil-producing countries were capped or protected, as for instance by the import quotas of the United States. Restrictions in the free trade of oil were a result of government intervention in the oil-importing countries, as already mentioned, not of oligopolistic actions by the oil corporations.

A few integrated multinational oil corporations, 'the seven sisters', are often alleged to have control of the flow of oil from the well to the gasoline station around the corner. But it is hardly fair to describe the oil market as an oligopoly. In many countries domestic oil corporations have long been competing with the seven giants. In the United States, where domestic oil production until recently has exceeded net imports of oil, the seven sisters in no way dominate the market. In Sweden, one of many industrial countries with no domestic oil production, the seven sisters have competition from national oil companies that either import crude for their refineries or import products. Russian oil and oil from American independents have long been competing with oil from the large integrated corporations as sources of supply in the spot market for small domestic oil companies.

The search for oil in distant countries under trying conditions was no string of success stories in the early part of the century as evidenced by the development in Latin America, especially Venezuela, before 1930. Oil companies going into the Middle East in the 1930s had a hard time finding partners willing to share the financial risks. But the great success of several consortia in this area attracted many companies to new oil regions opened up in the 1950s and later. In the early 1970s well over a hundred companies were participating in consortia searching for oil in the North Sea.

Petroleum and petroleum products dominate international trade in tonnage. By value they account for over one-third of international trade in minerals. By comparison other minerals are small.

The iron and steel industry, the aluminum industry and the copper industry are major examples of mineral processing industries. All differ from each other and from the petroleum industry in geographic location, business structure, pricing and raw material policy.

The iron and steel industry has traditionally been national in scope. Administered prices are typical for steelworks turning out a host of semi-manufactured products, the leading raw materials of the large engineering industry. Oligopoly under close observation from the government's anti-trust-agency is typical of the steel industry in most industrial countries. Steel corporations obtain their iron ore and coking coal from mining subsidiaries or buy them in the market on long-term or short-term contracts. American steel groups early integrated backwards and acquired domestic as well as foreign ore mines. The acquisition by Bethlehem Steel of iron ore mines in Chile before the First World War was an early example of a post-World-War II trend when American steel groups became involved in the development of new ore fields in Venezuela, Labrador, Liberia and elsewhere. American mining interests went abroad when it became obvious that the dominating high-grade ores near Lake Superior would soon be exhausted. The low-grade taconite ores were more expensive to mine.

The competitive position of domestic and foreign ores have shifted in the United States. The revolutionary increase in the size of ocean-going bulk carriers after 1955 tilted the scales in favor of foreign ores; the recent opening of a large lock in the Soo Canals between Lake Superior and Lake Michigan doubles the size of lake carriers (to more than 56 000 dwt) that bring taconite pellets from Lake Superior ports to the steel centers along the lower lakes which substantially improves the competitive situation of taconite. American mines abroad have been either fully owned by a steel subsidiary (e. g. the recently nationalized mines in Venezuela, owned by US Steel and Bethlehem) or operated as joint ventures with foreign mining companies (Bethlehem and Gränges in the Lamco-Project in Liberia).

The remarkably expansive Japanese steel industry has organized its ore (and coking coal) supply by writing long-term contracts with consortia incorporating a large number of companies. The Pilbara region of northern Western Australia, which is rapidly becoming the world's leading iron ore exporting district, may serve as an example. In a typical consortium some Australian companies, often with no previous experience in mining, provide finance and local strings, one or more American companies contribute technical, managerial and financial capacity and often one or more Japanese companies provide market connections and finance. The key to this development model is the long-term contract. More recently Australian mines have shipped iron ore and coking coal also to Europe.

The aluminium industry is global in scope and controlled by fewer corporations than the petroleum industry of the 1950s. Administered prices and oligopoly are typical. The industry is integrated from the bauxite mine to the aluminum refinery which produces raw materials for the engineering industry. Bauxite mines are primarily located in tropical regions, including northern Australia. Processing reduces weight and increases value. Bauxite in 1970 was

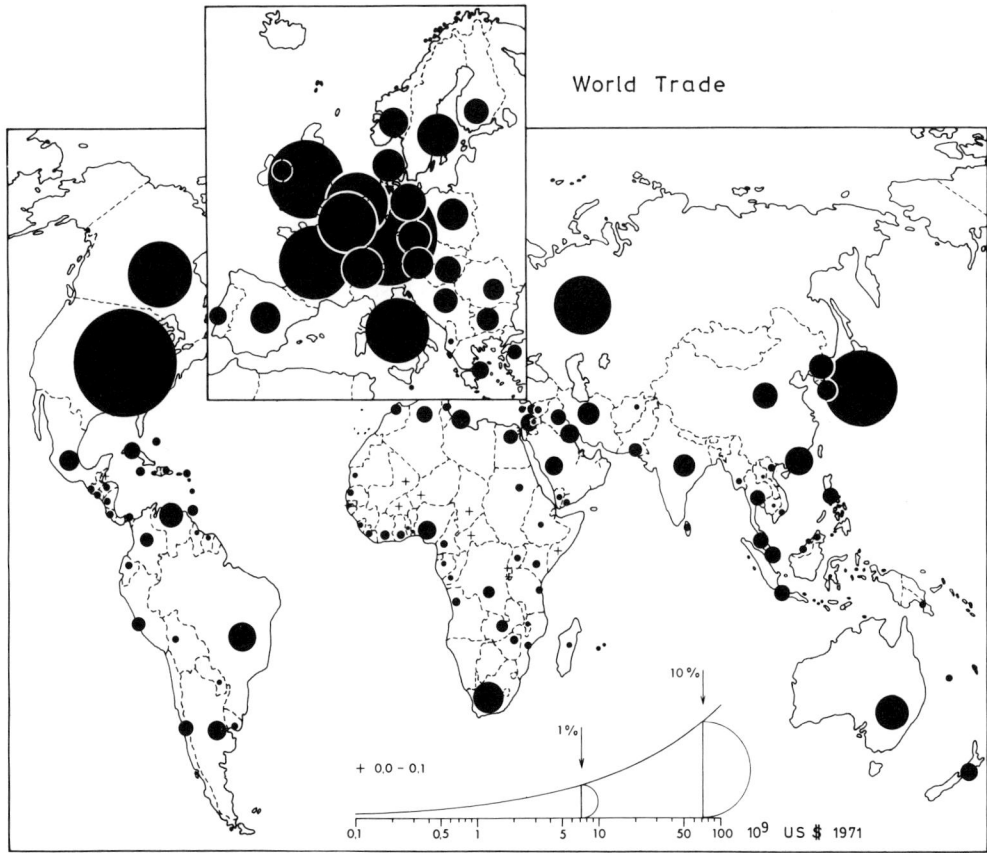

Fig. 41 – The map of world trade – exports plus imports for each country – serves as a base map for judging the two maps of mineral imports (US and Japan), figs 42 and 43.

some US $20 per ton and aluminum roughly $550 per ton. Aluminum groups are always looking for cheap electricity for their refineries. Natural gas fields and waterpower in areas remote from energy markets, ceteris paribus, provide the cheapest electricity. In spite of a huge water power potential in some LDCs, e. g. equatorial Africa, or large unexploited natural gas fields, e. g. in the Persian Gulf region, the aluminum groups have hesitated before making heavy investments in such unstable areas. Nationalizations and breaches of contracts in connection with the oil crisis have not made them less hesitant to take steps that from the point of view of pure economic geography would be obvious. The copper industry is also global in scope and controlled by a few large corporations. In addition, it has few buyers. In spite of this, prices are fluctuating strongly. In contrast to steel and aluminum, copper is a homogenous product. Copper bars and a few standardized products are sold at auctions in the London

Metal Exchange (LME), either directly from its warehouse (loco) or through futures (3 months). Only part of the world trade in copper occurs at LME prices, the so called 'world market price'.

International trade in minerals. Table 54 reveals that the 1970 international trade in minerals primarily was trade between the DCs. The LDCs with $^2/_3$ of the world's population and $^4/_7$ of its land surface accounted for less than $^1/_7$ of its imports of minerals and for $^4/_9$ of its exports. Almost $^3/_4$ of the exports of these SITC-groups from the LDCs were petroleum and petroleum products. The oil exports to an overwhelming extent originated in a few countries, most of them around the Persian Gulf. Measured by their GNP/capita some of these countries did not belong to the group of poor countries but were high up at the other end of the scale.

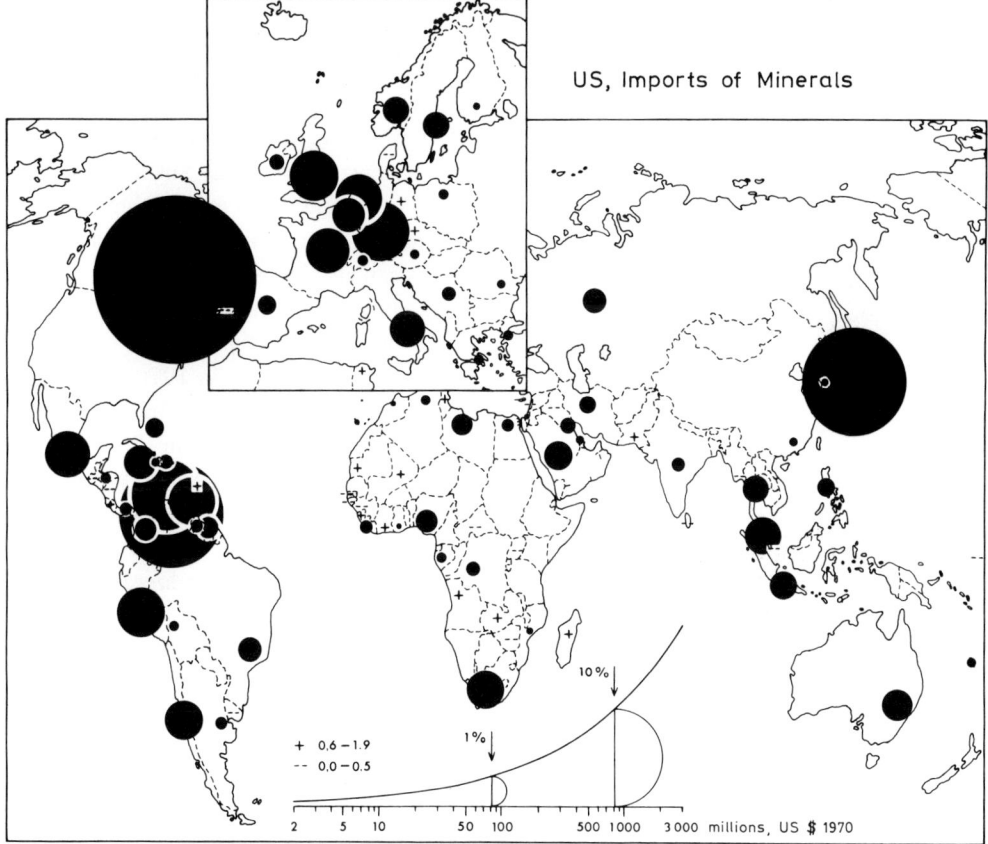

Fig. 42 – The imports of minerals into the United States is dominated by Canada, which accounts for more than Latin America and Africa combined. As shown by figs 42 and 43 and by table 54, international trade in minerals is primarily trade between industrial countries with significant amounts also coming from less developed countries.

The two maps showing the imports of minerals in 1970 to the United States and Japan reveal the same pattern. The two countries were almost equal buyers of minerals and together accounted for $^2/_7$ of the mineral imports into the DCs. The maps refer to the same SITC-groups as the table and were based on national trade statistics.

For comparison, maps were drawn of the populations and land surfaces of the countries in the world. The symbol scales on the four maps were chosen to facilitate comparison. The total symbol surface is the same on all maps which means that e. g. 0.1%, 1% or 10% of the world total is represented by the same symbol-sizes on all maps. This permits direct comparison of quantities expressed in US dollars, inhabitants or square kilometers.

United States in 1970 imported as much minerals from Canada as from the whole of Latin America and Africa combined. Japan accounted for more than the rest of Asia, including the Persian Gulf region. Several countries in Europe exported more minerals to the United States than Chile, a nation in which the 'exploitation' issue was hotly debated in the 1960s. The country in the fall of 1970 got a Marxist government[31] which was ousted in 1973. Like many other metals, copper can be obtained from low-grade ores in the DCs or from high-grade ores in the LDCs. In all its uses copper has substitutes. The bargaining power of the LDCs is therefore weak. Since many metals are classified as strategic, several DCs, among them the United States, have built stockpiles for an emergency. The stockpiles can also be used when a government finds that metal prices are getting out of hand. Actions by a large producer-consumer like the United States influences world market prices.

For Japan, the Persian Gulf region carried much weight in 1970, more than Anglo-America. But Japan is in an interesting geopolitical situation. In the perspective of a decade or two, three or four alternatives to the Persian Gulf may open up. The vast eastern regions of the Soviet Union contain much fossil fuel, primarily coal and natural gas but future finds of oil should not be left out. For this sparsely populated area Japan is a natural market. China is more and more coming into focus as a potential petroleum producer on a large scale. The domestic market for oil in China is tremendous but the country also needs foreign exchange to pay for her imports of foreign technology. Japan is the logical market for Chinese oil exports. The shelf off the coast in East Asia has very promising structures for oil and natural gas, some rather close to Japan, but the geopolitical situation is by no means clear and it may take a long time before dividing lines have eventually been drawn. The strategy adopted in the North Sea was for neighboring countries to first lease the tracts closest to the dividing line to tap pools stretching across the line. Indonesia is already an important supplier of oil to Japan and although a member of OPEC it is a major alternative to the so far dominating Persian Gulf basket. Its production is likely to continue to expand.

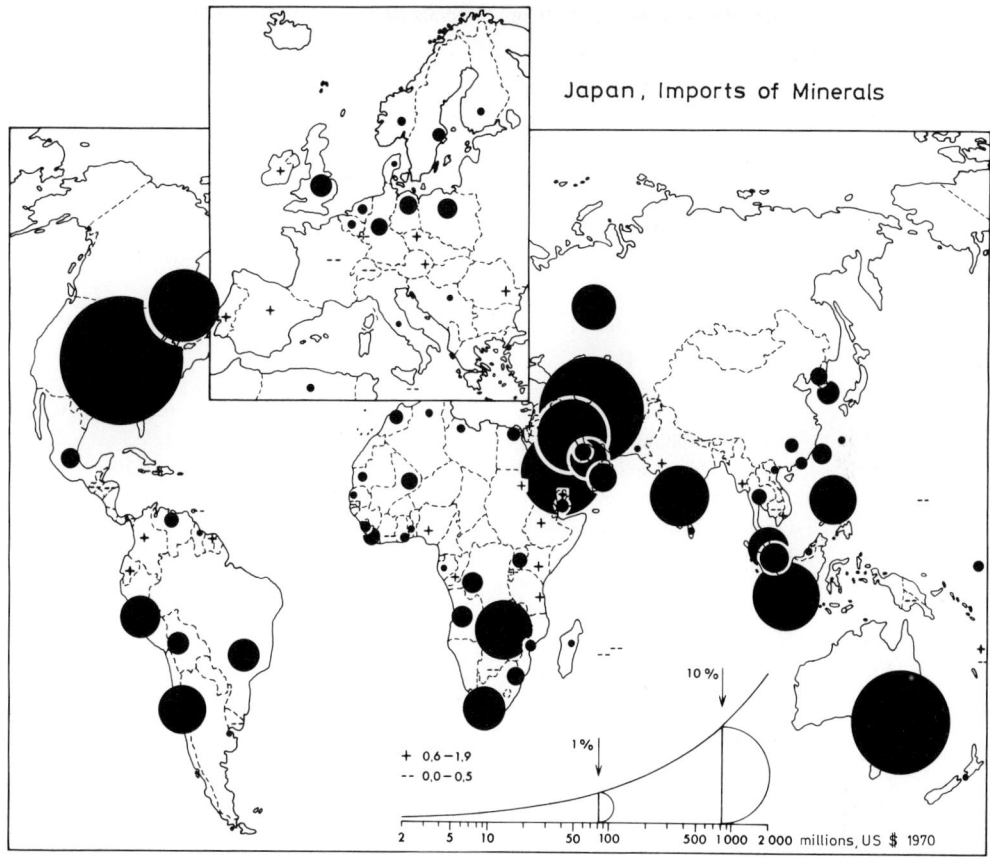

Fig. 43 – The most important supplier of minerals to Japan is North America but Australia is the most expansive supply region. For commodity after commodity it is taking the first position.

The single most important supplier of minerals to Japan in 1970 was the United States with Canada as another major exporter. Both countries shipped a vast spectrum of minerals to Japan. By comparison Latin America and Africa were rather small suppliers.

In the western Pacific region Australia with its large land surface and small population more and more emerges as the great storehouse of minerals for Japan, a great industrial nation with a small land base. In the Australian rush for minerals in the 1960s more high-grade iron ore was found than the world held in reserves in 1960. For bauxite the situation was similar. The list of minerals that Australia can supply is long. The rapid development of these mineral resources was made possible by Japanese long-term contracts. Consortia were formed with Australian, American or European and often also Japanese participation.

Tab. 54 – International Trade in Fuels and Minerals 1970
(million US $)

SITC	Commodity	Imports: DCs	LDCs	Exports: DCs	LDCs
271	Crude fertilizers	438	63	107	195
273	Stone, sand and gravel	309	34	253	15
274	Sulphur etc.	207	53	129	31
275	Natural abrasives	249	10	220	24
276	Other crude minerals	1 132	174	873	400
281	Iron ore, concentrates	3 041	26	1 279	1 012
282	Iron and steel scrap	1 003	144	870	41
283	Nonfer base metal ore, concentrates	3 266	96	1 617	1 010
284	Nonfer metal scrap	896	33	621	63
285	Silver and platinum ores	253	0	125	22
286	Uranium, thorium ore, concentrates	9	2	28	6
321	Coal, coke, briquettes	2 907	186	2 311	15
331	Crude petroleum, etc.	14 512	2 795	799	17 572
332	Petroleum products	5 478	1 442	3 767	3 175
341	Gas, natural and manufactured	710	108	625	150
671	Pig iron, ferroalloys etc.	1 075	59	744	108
672	Iron, steel primary forms	1 441	428	1 408	46
673	Iron and steel shapes	2 581	670	2 910	131
674	Iron, steel univ, plate, sheet	3 830	1 081	5 070	84
675	Iron, steel hoop, strip	522	94	703	1
676	Railway, rails etc. iron, steel	60	82	135	11
677	Iron, steel wire excl. wire rod	370	123	499	4
678	Iron, steel tubes, pipes etc.	1 486	505	2 344	48
679	Iron, steel castings unworked	145	41	287	1
681	Silver, platinum, etc.	809	16	560	68
682	Copper	5 229	381	3 065	2 697
683	Nickel	1 059	38	941	74
684	Aluminum	1 989	254	2 044	116
685	Lead	364	35	334	66
686	Zink	306	94	293	63
687	Tin	522	51	136	472
689	Non-fer base metals nes	487	25	396	83
	Total	56 685	9 143	35 493	28 084

Source: YITS 1972–73   SITC = Standard International Trade Classification

These consortia had the financial, managerial and technical capacity to start mining in remote areas with a harsh climate and no infrastructure.

Developments in Australia and Canada in the 1960s may show the way for future developments in the LDCs, assuming that they create acceptable rules of the game in their respective countries. Mining groups with financial, managerial and technical capacity have many areas to choose among, most of them in the DCs. Mining will hardly ever be a large industry expressed in employment but may make significant contributions to a country's exports. It may easily become the base of a much larger processing industry.

LDCs cannot afford to leave their mineral riches in the ground waiting for future temporary shortages and high world market prices. They must develop their economies now to provide employment for their rapidly growing populations. A blackmail strategy of the OPEC-model against the DCs is doomed. The DCs have the strongest cards: the market for mineral raw materials and the capacity of being self-sufficient with minerals from their own territories and from the seabed. They have by far the largest industrial production apparatus and the capacity of building large food surpluses. And last but not least, they have complete superiority in technical knowledge, research and development. The Marxist exploitation philosophy leads to confrontation and shrinking world trade. The liberal philosophy of buying where goods are cheapest and selling where the best price can be obtained or, with other words, free world trade instead of primitive barter arrangements will in the long run be best also for the LDCs who for most minerals enjoy large land rents. In a stable world without blackmailing in raw materials the LDCs would have good chances of attracting processing industries, maybe also engineering industries, which are much superior to mining in creating employment and high values added by industry.

Notes

[1] This does not imply that the LDCs, which overwhelmingly are tropical countries, in the foreseeable future can be expected to reach the per-capita-consumption of the DCs for essential minerals.

[2] The Law of the Sea Conference at Geneva in 1975 and New York in 1976, which continued the conference at Caracas in 1974, was called by the United Nations to draw up rules for the administration of the deep sea floor, a prerequisite for mining. This major item on the agenda had been left unresolved by the first two conferences at Geneva in 1958 and 1960 which drew up rules for the utilization of mineral deposits on the shelf. The DCs and the LDCs were reported to be at loggerheads over the issue. The DCs wanted an agreement, the LDCs who had no capacity to utilize the riches of the sea were in no hurry.

Almost all countries were ready to widen the historic 3-mile limit to 12 nautical miles. Within this limit coastal states should have full sovereignty. Within a zone out to 200 miles, coastal states should have economic control of fishing and mining. Some countries (US, Canada, Australia etc.) wanted the right to produce oil on the continental shelf even beyond 200 miles.

An international authority was suggested for the seabed beyond 200 miles. This authority should collect royalties for mining which would benefit all countries, even the landlocked ones.

[3] In the United States no less than 3.2 kg coal were used in 1899 to produce 1 kWh of electricity. In 1919 only 1.5 kg were used and in 1970 no more than 0.4 kg. In the first seven decades of this century the coal input for making 1 kWh thus was reduced almost 90 percent.

[4] The by-product coke oven was developed in Germany at the end of the 19th century and rapidly became popular with the expansion of the coal tar chemical industry. The old beehive ovens, in which the volatile matter was lost, soon were closed down. Thereby the geography of coke making was changed. The many small beehive ovens, typical weight-loss processes, were scattered over the coalfields. The large by-product coke ovens were built adjacent to the iron works which were located at interior waterways near the coal mines.

The by-product coke ovens were rapidly accepted in Germany with its world monopoly in the coal tar chemical industry but slowly in the United States which reached a prominent position in the chemical industry much later.

City gas has a low heating value compared with natural gas. In most towns the gasworks have been closed and the towns have been hooked on to the natural gas system. In regions without access to natural gas a gas fraction from oil refineries is fed into the pipe system. Also electricity is an attractive source of energy for city households.

[5] Even in absolute terms coal production declined in the 1960s (almost 70 million tons in oil equivalents), see also figure 12. In the same period the oil consumption of OECD-Europe increased from 199 to 620 million tons. The share of oil increased from 33 percent to 60 percent.

[6] Coal in 1973 accounted for 34% of West Germany's electricity production, brown coal for 32%, natural gas and oil each for almost 10%, water power for 9% and nuclear power for 5%.

[7] United Mine Workers (UMW) from 1920 was under the vigorous leadership of John L. Lewis, who played a prominent part in American politics and in 1935 became the President of the newly formed Congress of Industrial Organization. CIO recommended unions to organize according to the industry principle (one union only at each place of work) while the older top organization, American Federation of Labor, was dominated by unions organized according to the occupation principle. The two organizations were merged in 1955 to AFL-CIO.

[8] The Soviet record worker Alexei Stakhanov in Donbass in the 1930s mined 102 tons in a six-hour-shift against a norm of 7 tons. But these statistics are not comparable. The American production is divided by all those who work in the mine. The Russian data refer to one man who mines the coal.

[9] A new change in the Soviet energy policy was reported in the fall of 1974 which was made public with the new five-year plan (1976–81). With open pits, electricity in base load stations can be produced almost as cheaply with coal as with oil or natural gas. Petroleum would then be reserved

10 The new industry was soon dominated by Standard Oil, a Cleveland company established in 1872. Under John D. Rockefeller this company bought a lot of firms and acquired control of pipelines and refineries. By 1878 Standard Oil controlled 90 percent of the American oil market.

for stand-by facilities taking care of the peak demand. Petroleum plants can be started more quickly. Oil and natural gas are more attractive than coal for exports and for the chemical industry.

The American antitrust legislation, the Sherman Act of 1890, had the oil trust as one of its major objectives. The Standard Oil Trust was formally dissolved in 1892 after a decision in an Ohio court. The organization continued to operate and was reorganized in 1899 as a holding company in New Jersey. But in 1911 the Supreme Court declared that the trust violated the federal laws and ordered Standard Oil to be broken up into smaller corporations.

11 Geopolitically, the stormy Barents Sea is a sensitive area to the Soviets, who have the largest part of their huge fleet of nuclear submarines — equipped with intercontinental robots, which can be launched from a submarine position — stationed in Murmansk. The submarines can easily be spotted from airplanes in the shallow waters of the shelf but can without difficulty hide in the deep waters of the Norwegian Sea.

12 Before the Iraq Petroleum Co (IPC) was nationalized in 1972, BP, Shell, Compagnie Française des Pétroles and a subsidiary of two American corporations (Exxon and Mobil) each owned ¼ of 95% of the shares and the original organizer, C. S. Gulbenkian, 5%.

13 Saudi Arabia with some 7 million inhabitants is three times the size of France. At the time of the oil crisis, it was a few months away from becoming the largest oil producer in the world. The oil corporations had published projections pointing at a production of 1 000 million tons a year by 1980. Shipowners had placed orders for giant tankers to carry this oil to the United States, Western Europe and Japan. The four-fold increase in oil price of late 1973 meant a ten-fold increase in state revenue per barrel.

Within OPEC, Saudi Arabia is said to have exercised a restraining influence on prices. Thanks to exceptionally favorable geological structures, Arabia has exceedingly low production costs. Before the oil crisis, Aramco enjoyed a high land rent which was divided between the state and the oil company. After the crisis the state levied the very stiff OPEC monopoly tax on top of its share of the land rent. The consuming countries could be expected to get together to develop alternative sources of energy, to help develop new petroleum fields outside the OPEC countries, and to make investments in energy-saving techniques. Such cooperation was inaugurated with the Washington Conference in February 1974 and the establishment of the International Energy Agency. Saudi Arabia does not want to risk that its oil market disappears within 10 years; it wants to sell oil for 50 or 100 years, for which the country has known reserves.

14 Orenburg was discovered in 1967 and a 3 200 km pipeline (140 cm) will be drawn to Uzhgorod near the Czechoslovakian and Hungarian borders. From these two Comecon-countries gas will be piped to Poland, Bulgaria and Yugoslavia but also to Austria, southern Germany, northern Italy and France. Investments in recent years have been concentrated to Orenburg which is better located than the Siberian gas fields.

The Soviet Union exported gas to Poland for the first time 1966, to Czechoslovakia 1967, to Austria 1968, to West Germany 1973, and to Finland 1974.

15 Armand Hammer, acting for the American oil group Occidental Petroleum, in the spring of 1973 signed an agreement of principles with the Soviet Union about delivery of petrochemical plants and so on during a twenty year period at a value of 8 billion dollars. According to the first contract, signed in June, 1974, the Americans are going to build, backed by credits from the Export-Import-Bank, four ammonia factories at Togliatti with a combined capacity of 1.8 million tons. They will be paid with ammonia shipped to the United States, one way of exporting natural gas in refined form. Gas, as already mentioned, is becoming a scarce commodity in the United States. The same company built the first Soviet ammonia factories in 1932—33.

Dr. Hammer as a green physician went to the Soviet Union in 1921 to organize the supply of food to starving people in the Urals, where his attention was drawn to asbestos deposits which were no longer being utilized. Hammer after a conversation with Lenin about exchanging asbestos for food grains was given the concession to mine the mineral in 1921. The Soviet leaders at this time wanted foreign capital help develop the raw material reserves of the country. The young American doctor should act as an icebreaker and open up economic relations with the outside world. After some more small concessions to foreign interests this development came to a standstill. Hammer, who made a career in American industry, continued to have good relations with the Soviet Union. He has also made original contributions in other parts of the world, e. g. developed large irrigation projects in the Sahara (at Kufra in the Libyan desert) and obtained oil from shale in Colorado.

16 As Einstein showed in 1905 in the most famous of equations, $E = mc^2$, matter may theoretically be changed into energy and the reverse. The equation says that the energy (E) of a system equals its mass (m) times the squared speed of light (c^2).

17 Wood represents some 50 years of accumulated sunshine and the fossil fuels were formed during long geological periods. The utilization of wood means tapping from the 'plant bank' into which deposits are continuously being made. Withdrawals should, however, be kept within the limits set by deposits if serious problems with capital reduction and soil erosion are to be avoided.

The 'mineral bank' is different. Withdrawals of fossil fuels and metals from this bank are not, as a rule, offset by deposits. This metaphor should not be carried too far, however, since it may easily lead us astray. The withdrawals from the mineral bank are made at a cost and the price will go up when demand exceeds supply according to basic rules of economics. And this is not the way a bank functions. High prices will encourage mineral saving methods and the search for substitutes. In addition, high prices will make low-grade deposits available changing them from resources to reserves (fig. 1). However, people with no training in economics and geology feel that they have a serious moral dilemma with the mineral bank. How much of the capital should be utilized by the present generation and how much should be left to future generations? How much should be utilized in one country and how much in other countries?

It is easy to show that man, by developing an energy and material concentrated technique, has adapted himself to changed conditions. Compare for instance the amount of coal equivalents utilized to produce 1 kWh of energy in 1710 and in 1975, the amount of timber used to produce 1 m^3 of dwelling space and so on. Table 4 indicates that mankind is not doomed to cataclysm by exhausted energy reserves. But the energy problem can not easily be swept away with a reference to man's technical adaptability. One pre-condition is a rapid decline in the growth rate of humanity, preferably to zero growth within relatively few decades. Otherwise environmental deterioration and social strains may lead to cataclysm even before energy has contributed any acute problems.

18 LD was developed 1949–52 at Linz and Donawitz, cities in Austria with steelworks. Kaldo was named for Kalling and Domnarvet. B. Kalling was professor of ferrous metallurgy at the Royal Institute of Technology in Stockholm; Domnarfvet is a major Swedish steelworks.

19 The Petsamo ore with a mine at Kotosjoki near present Nikel was developed after 1934 by Mond Nickel, a subsidiary of Inco. Nikel has railroad to Pechenga and Murmansk.

20 In Katanga cobalt, uranium, zinc, lead, silver, germanium and cadmium are associated with copper while Zaire's production of gold, tantalum, columbium, beryllium, tungsten, bismuth and rare earth metals occurs in connection with tin mining in the eastern mining province.

The Belgian Union Miniére, formed at the beginning of the century to develop the rich ores of Katanga (now Shaba), in 1969 reached an agreement with the state mining company of Zaire, Générales Congolaise des Mines (Gécomines). The center of copper-cobalt-mining is in the west near Kolwezi with three open pit and a shaft mine. Two electrometallurgical plants (Luilu and Shituru) based on water power smelt copper and cobalt. They receive sulfide and oxide concentrates from concentration plants at Kamoto and Kolwezi. The company mines some 10 million tons of ore a year.

21 At Sudbury Inco and Falconbridge are equally important as cobalt producers. The latter company obtains its cobalt at its nickel-copperrefinery at Kristiansand, Norway.

22 The Ytterby feldspar mine on Resarö 3 km north of Vaxholm in the archipelago of Stockholm played an important role in the naming of the rare-earth metals. John Gadolin of Turku Åbo University, J. J. Berzelius and C. G. Mosander made important contributions to the identification of these elements of which gadolinium (Gd) is named for Gadolin, terbium (Tb), erbium (Er), and ytterbium (Yb) for Ytterby, holmium (Ho) for Stockholm, Scandium for (southern) Scandinavia, and thulium (Tm) for Thule.

23 Technically, cement is any of various soft, sticky substances that dry hard or stonelike, used for making things adhere. As a construction material, portland cement is now almost the only kind of cement being used; in common usage portland cement and cement are synonyms.
24 The engineering firm F. L. Smidth of Copenhagen designed the first rotary kiln for an Ålborg plant in 1898. It has since built over 1 000 such kilns for cement factories around the world. Some 95% of its production is for exports.
25 Competing firms use the Hartford-Empire machines. This pure machine company used to lease their machines to glass firms on conditions that were thought to be restraining on competition. After court decisions in Toledo and in the Supreme Court, an agreement was reached in the 1940s that Hartford-Empire was to sell their machines instead of leasing them and to sell patent licenses for machines against royalty. Owens-Illinois now uses machines from Hartford-Empire in addition to its own machines and a mix of machines is common also in other companies.
26 The dramatic increases in world market prices for food between 1972 and 1974 were caused both at the input and at the output end. At the output end, the large Soviet purchases of grain in 1972 swept away the American reserves that long had acted as a buffer. Grain prices increased rapidly, both in the domestic market of the United States and in the world market. At the input end, the major cost items of the grain farmer went up rapidly, primarily energy and fertilizers. As already shown, the most important fertilizer, nitrogen, from an economic point of view is almost synonymous with energy.
27 The negative logarithm of the hydrogen-ion concentration in gram equivalents per liter. In a scale from O (acid) to 14 (alkaline), neutrality is represented by 7 (pure water at 25°C). Soils vary in pH from 3.5 to 9 but the range favorable for ordinary crops is from 6.0 to 7.5.
28 Power is an attribute of the government who promulgates and enforces laws and who enters into agreements with foreign countries, binding to all her citizens, and who may wage war. Individuals or organized groups of individuals such as business firms, trade unions or golf clubs have no such rights but a large business firm may have much economic influence (sometimes called power) and a great philosopher may have much social or religious influence (also called power). The concept is here used exclusively in the first sense.
29 In hindsight, the success of OPEC in 1973 may turn out as a Pyrrhic victory. The oil-exporting countries lost large parts of their markets when exploration for oil was rushed in other parts of the world, nuclear power projects got a hike and energy saving techniques were pushed. The cost to the oil-importing countries were tremendous: two-digit inflation, the largest postwar unemployment and a precarious balance of payment situation.
30 OPEC became a multistate monopoly when unilaterally it raised the price of oil above the global marginal cost of producing oil. Before that, the profits accruing to the exporting countries, being divided between the state and the oil-producing company, were land rent bestowed on these countries by nature. After that, the oil-exporting nations made the citizens of the oil-importing countries pay an additional monopoly tax.

Variations in land rent from one oil district to another is a matter of indifference to the oil consumer but are of great importance to the oil-producing countries and regions. Monopoly taxes on the other hand, whatever their size, are unacceptable for the consumer and for the importing countries. No modern country can be expected in the long run to accept that its citizens are forced to pay tax to another country. Such tributes were sometimes paid in the past. In the distant past, countries in Asia paid tributes to the Emperor of China and in the more recent past countries of Europe, including Sweden, paid tributes to the Sultan of Morocco.

With a multistate monopoly among leading exporting countries the supply of oil becomes a concern of political geography in the importing countries and not primarily an affair of economic geography.
31 deVylder in his book on Chile during the Allende government shows that two arguments were current in the domestic debate in the 1960s: the American copper corporations paid too little for the metal and they did not push the Chilean production suffiently to hold the country's share of world production.

To prove the first point Chilean copper prices were compared with the spot prices at the London Metal Exchange, which were higher. A more relevant comparison would have been prices paid for the much larger American production, which were lower. The American corporations thus charged a higher price for low-cost Chilean copper (average metal content in the ore some 3%) than for high-cost American copper (0,6% metal content). United States paid more for imported copper than for domestic. From 1968 the Chilean copper was sold on the London Metal Exchange and within a few years Chilean prices coincided with LME-prices. It is doubtful if Chile in the long run will be better off with the new arrangement.

To prove the second point Chilean economic writers quoted by de Vylder showed Chile's percentage of world production at two points without mentioning that the decline in Chile's percentage was a result of the rapid postwar expansion of the Soviet copper production. Chile in the 1950s and 60s had a much more rapid expansion than the United States and roughly the same as Zambia, its main competitor.

Measure Prefixes

T	tera	=	10^{12}	d	deci	=	10^{-1}
G	giga	=	10^{9}	c	centi	=	10^{-2}
M	mega	=	10^{6}	m	milli	=	10^{-3}
k	kilo	=	10^{3}	μ	mikro	=	10^{-6}
h	hekto	=	10^{2}	n	nano	=	10^{-9}
da	deka	=	10^{1}	p	piko	=	10^{-12}
				f	femto	=	10^{-15}

Capacity

1 horsepower ≎ 736 watt (W) ≎ 0.736 kilowatt (kW)

Energy

1 therm ≎ 29.3 kilowatthours (kWh) ≎ 100 000 British thermal units (Btu)
1 Btu ≎ 0.293 watthours (Wh) ≎ 1 054 watt seconds (Ws)
1 Q ≎ 10^{18} Btu ≎ 293 · 10^{12} kWh
1 megacalorie (Mcal) ≎ 1.163 kWh ≎ 1 hermie
1 kWh ≎ 3 411 Btu ≎ 859.6 kilo calories (kcal)
1 calorie (cal) ≎ 4.19 Ws
1 watt-second (Ws) ≎ 1 joule (J)
1 electron-volt (EV) ≎ 1.6 · 10^{-19} Ws
1 ton coal ≎ 7 600 kWh
1 m^3 crude oil ≎ 10 120 kWh
1 kg uranium ≎ 23 000 000 kWh (theoretic maximum)
1 kg uranium oxide (U_3O_8) ≎ 180 000 kWh (thermic reactor)
1 kg uranium oxide (U_3O_8) ≎ 12 600 000 kWh (breeder reactor)
1 m^3 natural gas ≎ 9.3 kWh
1 m^3 liquefied natural gas (LNG) ≎ 600 m^3 natural gas
1 ton crude oil ≎ 1.16^3 crude oil ≎ 11 740 kWh
1 barrel (bbl) crude oil ≎ 0.159 m^3 crude oil ≎ 1 610 kWh

Short Bibliography

Georgius Agricola, De Re Metallica, 1556. Transl. by Herbert C. Hoover & L. H. Hoover 1912, The Mining Magazine (London).
Donald A. Brobst & Walden P. Pratt (eds.), United States Mineral Resources (Washington: GPO, 1973). US Geological Survey Prof. Paper 820.
Peter Burton, Trends in World Shipping with Particular Reference to Large Crude-Oil Carriers (La Spezia: NATO, 1974).
Geoffrey Chandler, "Energy" in Inglis. Energy: From Surplus to Scarcity? (Barking: ASP, 1974).
P. E. Cloud, jr., Resources and Man (San Francisco: Freeman, 1969).
Earl Cook, "The Flow of Energy in an Industrial Society", Scientific American 1971.
Earl Cook, Man, Energy, Society (San Francisco: Freeman, 1976).
P. F. Corbett, "Natural Gas – Growth of a World Market" in Inglis (ed.), Energy: From Surplus to Scarcity? (Barking: ASP, 1974).
Herman E. Daly, (ed.), Toward a Steady-State Economy (San Francisco: Freeman, 1973).
H. P. Drewry (Shipping consultants) Ltd., London, World Iron Ore Trade and Shipping Requirements, London 1973.
K. O. Emery, "Geological Aspects of Sea-floor Sovereignity", in Lewis M. Alexander, The Law of the Sea: Offshore Boundaries and Zones (Columbus: Ohio Univ Press, 1967).
Fearnley & Egers Chartering Co. World Bulk Trades 1973 (Oslo, 1974).
R. Geipel, Industriegeographie als Einführung in die Arbeitswelt (Braunschweig: Westermann, 1969).
G. J. S. & M. H. Govett, "The Concept and Measurement of Mineral Reserves and Resources", Resources Policy, Vol. 1, 1974.
N. B. Guyol, Energy in the Perspective of Geography (Englewood Cliffs, Prentice-Hall, 1971).
Allen L. Hammond, William D. Metz & Thomas H. Maugh II, Energy and the Future (Washington: Amer Ass for the Advancement of Science, 1973).
Orris C. Herfindahl, "Some Problems in the Exploitation of manganese Nodules". in Lewis M. Alexander (ed.), Needs and Interests of Developing Countries (Kingston: Univ. of Rh I, 1973, Proceedings of the 7th Annual Conference of the Law of the Sea Institute, 1972).
R. M. Highsmith & J. G. Jensen, Geography of Commodity Production (New York: Lippincott, 1958).
Ivar Högbom, Mineral Production: A Study in Trend and Geographical Displacement, Ingenjörsvetenskapsakademien, Proceedings, No. 117, Stockholm 1932.
M. Hollander & M. K. Simmons (eds.), Annual Review of Energy, Vol. 1 (Palo Alto, 1976).
M. King Hubbert, "Energy Resources" in Resources and Man (San Francisco: Freeman, 1969). National Academy of Sciences Committee on Resources and Man.
–, "The Energy Resources of the Earth", Scientific American, No 3, 1971.
Johannes Humlum, Kulturgeografisk Atlas, II Tekstbind, 7th ed., (Copenhagen: Gyldendahl, 1971).
Ilmari Hustich, Världens klyftor och broar (Helsingfors: Schildt, 1973).
K. A. D. Inglis (ed.), Energy: From Surplus to Scarcity? (Barking: Applied Science Publishers, 1974).
C. F. Jones & G. G. Darkenwald, Economic Geography, 3rd ed., (New York: Macmillan, 1965).
S. A. Lawrence, International Sea Transport: The Years Ahead. Lexington, Mass. 1972.
Hans Linneman, An Econometric Study of International Trade Flows. Amsterdam 1966.
Paul E. Lydolph, Geography of the USSR, 2nd ed., (New York: Wiley, 1970).
T. R. Malthus, An Essay on the Principle of Population, 9th ed., (London: Reeves and Turner, 1888).

Brian Mason, Principles of Geochemistry, 2nd ed., (New York: Wiley, 1958).

D. H. and D. L. Meadows, J. Randers & W. W. Behrens, III, The Limits to Growth (New York: Universe, 1972).

H. W. Menard, Marine Geology of the Pacific (New York: McGraw-Hill, 1964).

J. L. Mero, The Mineral Resources of the Sea (New York: Elsevier, 1965).

—, "Alternatives for Mineral Exploitation" in L. M. Alexander (ed.), The Future of the Sea's Resources (Kingston: Univ. of Rhode Island, 1968), Proceeding of the Law of the Sea Institute, 1967.

Walter E. Morrow, Jr., "Solar Energy: Its Time is Near", Technology Review, vol 76, 1973.

National Academy of Sciences, Mineral Resources and the Environment (Washington: NAS, 1975).

National Petroleum Council, US Energy Outlook: An Initial Appraisal 1971–1985 (Washington: NPC, 1971).

Peter R. Odell, Oil and World Power: A Geographical Interpretation, 3rd ed., (Harmondsworth: Penguin, 1975).

—, "Europe and the Cost of Energy: Nuclear Power or Oil and Gas?" Energy Policy, Vol. 3, 1976.

Raymond L. Parker, Data of Geochemistry (Washington: GPO, 1967), Geological Survey, Prof. Paper 440-D.

W. N. Peach & J. A. Constantin, Zimmermann's World Resources and Industries, 3rd ed. (New York, Harper & Row, 1972).

P. H. Phizackerley & L. O. Scott, "Major Tar Sand Deposits of the World", Proceedings 7th World Petroleum Congress, Mexico 1967.

V. V. Pokshishevskij (ed.), Sowjetunion, Regionale ökonomische Geographie (Leipzig: Gotha, 1967).

'The Rasmussen Report' – US Nuclear Regulatory Commission, Reactor Safety Study: An Assessment of Accident Risks in U.S. Commercial Nuclear Power Plants. Rep. No. WASH-1400 (NUREG 75/104) (Washington: GPO, 1975)

Rolf Schniotalle, Der Braunkohlenbergbau in der Bundesrepublik Deutschland (Wiesbaden: Steiner, 1971).

Brian J. Skinner, Earth Resources, 2nd ed., (Englewood Cliffs: Prentice-Hall, 1976).

– & K. K. Turekian, Man and the Ocean (Englewood Cliffs: Prentice-Hall, 1973).

Stanford Research Institute, Patterns of Energy Consumption in the United States (Washington: GPO, 1972).

Carol & John Steinhart, Energy: Sources, Use and Role in Human Affairs (North Scituate: Duxbury Press, 1974).

E. N. Tiratsoo, Oil Fields of the World (Beaconsfield: Scientific Press, 1973).

A. R. Ubbelohde, Man and Energy (New York: Braziller, 1955).

K. Warren, Mineral Resources (New York: Wiley, 1973).

A. Zimm, Industriegeographie der Sowjetunion (Berlin: DVW, 1963).

Erich W. Zimmermann, World Resources and Industries, 2nd ed., (New York: Harper, 1951).

Statistical Sources

Commodity Research Bureau, Inc., Commodity Yearbook 1974. New York, 1974.
Fearnley & Egers Chartering Co. Ltd., World Bulk Trades 1973, Oslo, 1974.
Mining Annual Review 1974, London 1974.
Metallgesellschaft AG: Metallstatistik 1963–1973, Frankfurt a. M., 1974.
Minerals Yearbook, 1970, Vol. 1, (Washington: 1971) Bureau of Mines, MY
Norodnoe Chosjajstvo USSR v 1972 g.
G. Sundbärg, Aperçus statistique internationaux, Vol. 10 (Stockholm, 1906).
UN, Yearbook of International Trade Statistics 1972–73, YITS.
UN, Statistical Yearbook, Annual, SY.
UN, World Energy Supplies 1969–1972. New York, 1974.
Weltmontanstatistik: Die Versorgung der Weltwirtschaft mit Bergwerkserzeugnissen, I–IV, M. Meisner (ed.) (Stuttgart, 1925–1939).

Index

Abadan, 61, 62
Aberdeen, 56
Abrasives, 203
—, natural, 205
—, synthetic, 203, 205
Abqaiq, 64
Acari mine, 116
Achinsk, 168
Acid soils, 203
Acreage restriction, 197
Adelman, M. A., 209
Adobe, 185
Administered prices, 137, 211
AEC, Atomic Energy Commission (US), 85, 134, 171
Aerospace industry, 101
AFL, American Federation of Labor, 219
AFL-CIO, 219
AGR, Advanced Gas Reactor, 88
Age of Electricity, 137
Agricola, Georg, 107
Ålborg, 180
Alcan, Aluminum Company of Canada, 166, 174
Alcoa, Aluminum Company of America, 164, 166, 170, 174, 175
Aldan River, 160
Alfvén, Hannes, 94
Alkaline soils, 203
Allard Lake, 177
All-electric society, 97
Allende, S., 141, 222
Alloy metals, 127
Almadén, 156
Almelo, 85
Alnico, 130
Alted, Aluminum Limited, 166
Alumina, aluminum oxide, 164, 171, 205
Alumina plant, 39
Alumina works, world, 172–73
Aluminum, 10, 73, 97, 127, 136, 163–175, 210
—, cheap electricity, 212
—, consumption, 166
—, corporations, 166
—, economics, 165

Aluminium, density, 175
—, price, 163, 212
—, production, 167, 169
—, uses, 165
Aluminum cable, 165
Aluminum cartel, international, 166
Aluminum smelters, world, 172–73
Aluminium Industrie AG, 164
Alunite, 174
Alusuisse, 164, 166, 175
American independents, 210
American Metal Climax, 143
American Smelting and Refining, 139
Amino acids, 194
Amherst, 179
Ammonia, 97, 220
Ammonium nitrate, 196
Ammonium phosphate, 196
Ammonium sulfate, 196
Amos, 184
Amuay Bay, 53
Anaconda Corporation, 138
Anglesite, 154
Anglo-American Corporation, 143
Anglo-Iranian Oil Company, 61
Anglo-Persian Oil Company, 61
Anshan, 123
Anthracite, 20, 25, 30, 31, 36, 72
Anti-trust authorities, US, 166, 187, 211
Anti-trust legislation, 166, 208, 220
Antimony, 132
—, world production, 132
Anvil, 154
Apatite, 168, 197, 198
Apparel industry, 31
Aqaba, 200
Arad, 200
Arabian Gulf, see the Gulf
Arithmetic progression, 11
Aramco, Arabian American Oil Co, 63, 64, 220
Arctic Ugol, 30
Arkalyk, 168
Arnhem Land's Aboriginal Reserve, 175
Arrhenius, Svante, 86
Aruba, 53
Asahan River, 174

Index

Arzew, 66
Asbestos, 177, 184, 220
ASEA-Atom, 90
Ashdod, 200
Ashes, 201
Ashland, 44
Aspdin, Joseph, 179
Asphalt, 41
Asphalt lake, 52
Atacama Desert, 195
Athabasca tar sands, 191
Atom, 81, 82
Atom bomb, 80, 85, 86
Atomic Age, 80
Atomic Energy Authority, UK, 102
Atomic number, 81
AT&T, American Telephone & Telegraph, 139
Auger mine, 22

Babbitt, metal, 147
Bakony Forest, 168
Baku, 43, 57
Baku-Batumi pipeline, 57
Bakwanga, 204
Ball clay, 189
Ballas, 204
Ballast, cement bags, 181, 184
Baltic Sea, 179
Baltimore, 114
Balucan, 124
Bandar Mashur, 62
Bandar Shahpur, 192
Bangka, 147
Bangui Magnetic Anomaly, 126
Baniyas, 63
Barcelona, 201
Barents Sea, 220
Bari, 168
Barium, 10, 82, 84, 156
Barre, 178
Barter, 217
Basalt, 178
Base load, 30
Base metals, 136–157
Basic open-hearth, 198
Bass Strait, 69
Bathurst, 150
Batumi, 57
Bauxite, 7, 39, 164, 170, 174, 177

Bauxite, international trade, 170, 171, 174
–, price, 211–212
–, production, 167
Bay of Fundy, 95
Bayer, K.J., 164
Bayer process, 164, 165
Beaufort Sea, 75
Beaumont, Elie de
Bedfort, 178
Beehive ovens, 219
Beira, 143
Beira-Benguela Railroad, 142, 143
Bejaia, 66
Bell System, 139
Belovo, 152
Ben Guerir, 199
Beneficiation, iron ore, 109
Bentonite, 189
Berg Aukas, 134
Bergius, Friedrich, 74
Bergman, Torben, 107
Bergslagen, 116
Berlin, 189
Beryllium, 221
Berzelius, J.J., 221
Bessemer, Henry Sir, 32
Bethlehem Steel, 115, 211
BHP, Broken Hill Proprietary Co., 125, 152
Biafra, 67
Bilbao, 120
Billiton, 147
Billiton N.V., 142, 174
Bimetallism, 161
Bingham, 137, 138, 161
Biosphere, 192
Birkeland, K., 196
Birmingham, 112
BIS, Bank for International Settlements, 160
Bismuth, 221
Bituminous coal, see coal
Black Country, 25
Blackmail strategy, 217
Blast furnace, 26, 27, 36, 109, 179
Blister copper, 137
Bog ore, 109
Bogutin, 131
Bohai, 69
Bohr, Niels, 80
Boksitogorsk, 168
Bomi Hills, 126

Bone meal, 197
Bones, 198
Bong, 126
Bonny Islands, 67
Bor, 146
Borax, 205
Boric acid, 205
Boron, 193 205–206
Boron carbide, 205, 206
Bort, 203
Boryslav, 60
Borralha, 132
Bosch, Carl, 196
Bou Azzer, 135
Bou Kraa, 199
Bougainville, 145
BP, British Petroleum Company, 62, 64, 67, 207, 220
Brass, 133, 137, 149
Bratsk, 168
Breeder reactor, 15, 82, 86, 91, 92, 103, 105
Bretton Woods, 161
Brick, 178
Brignoles, 167
Brine, 175
Briquette, 23, 30
Britain, workshop of the world, 95
British Gas Corporation, 77
Broken Hill, 150, 152, 154
Broken Hill South, 152
Bronze, 133, 137, 147
Bronze Age, 137
Broström Group, 121
Brown, Ch., 23
Brown coal, 20, 166, 201, 219
–, aluminum, 171
Brucite, 175
Buchanan, 126
Bulk cargo, transport revolution, 110
Bulk carriers, 112
–, ocean-going, 211
Bullion, 158
Bunker coal, 26, 38
Bunker oil, 46
Bunker station, 26
Burgan, 64
Burmah Oil, 61, 69
Butane, 72
Butterworth, 147
Butte, 138

BWR, Boiling Water Reactor, 84, 89, 90, 92, 104
By-product coke oven, 219

Cadmium, 156, 221
Calcium, 10, 193, 203
Calcium cycle, 192
Calder Hall, 87
Caliche, 195
California gold rush, 158
Caltanisetta, 191
Camarines Norte, 124
CANDU, 85, 88
Cannel coal, 20
Canton trade, 189
Capital and technology, Belgian, 120
–, French, 120
Carat, 204
Carbamide, 196
Carbon, 10
Carbon dioxide, 86
Carbon steel, 127, 135
Carbonados, 204
Carlsbad, 202
Carnallite, 175, 201
Carnotite, 82, 134
Carol Lake, 114
Carrara, 179
Casablanca, 199
Cassiterite, 147
Catastrophism, 108
Catavi, 149
Caustic soda, 164, 170
CEGB, Central Electricity Generating Board (UK), 88
Cement, 168, 178, 179–84, 222
–, international trade, 183
–, production, 182, 183
Cement block, 178
Center – periphery, VIII
Central plan economies, 122, 197
Cerium, 133, 135
Cerro, 141
Cerro Bolivar, 115
Cerro de Pasco, 142, 152, 162
Cerussite, 154
Cgs-system, 158
Chagan-Uzum, 156
Chalcocite, 137
Chalcopyrite, 137

234 Index

Chalk, 180
Chapman, D., 100
Charcoal furnaces, 26
Charleroi, 186
Chelyabinsk, 152
Chemical identity, 81
Chemical industry, 201
Cherepovets, 117, 123
Chernigov, 60
Cherrapunji, 96
Chiatura, 127
Chile saltpeter, 194, 195–196
Chimkent, 152
Chin Ling, 133
China, Emperor of, 222
Chingwangtao, 68
Chlorine, 10
Chlorophyll, 194
Chromite, 129
–, world production, 129
Chromium, 7, 10, 127, 129, 156
Chromtay, 129
Chuquicamata, 133, 137, 140
Churchill, Winston, 80
CIA, Central Intelligence Agency (US), 128, 160
Cia Vale do Rio Doce, 115
Cie Générale d'Electricité, 89
Cinnabar, 156
CIO, Congress of Industrial Organization, 219
Cipec, Council of Copper Exporting Countries, 138
City gas, 20
Ciudad Guayana, 115
Clay, 180, 185, 205
Clean Air Act, 32; 163
Clean oil, 46
Cleveland, 119
Climax, 132, 133
Clinch River, 93
Clinker, cement, 180
Club of Rome, 103
Coal, 17, 18, 19–41, 72, 73, 74, 86, 91, 170
Coal cycle, energy, 86
Coal, Donbass, 120
–, electricity, 219
–, international trade, 41
–, markets for, 31–32, 35, 38, 41
–, origin, 19–20
–, protection, Western Europe, 208

Coal, reserves, 19, 35, 40
–, types of, 20
Coal crisis, Japan, 37
Coal mines, productivity, 25, 31, 41
Coal mining, risks, 104
Coal slurry, 40
Coal tar, chemical industry, 219
Cobalt, 127, 128, 134, 186, 193, 221
–, radio-active isotope, 135
–, world production, 134
CODELCO, Corporación de Cobre (Chile), 141
Coeur d'Alene, 154
Coke, 20, 27, 31
Coke oven, gas, 186, 196, 219
Coking coal, 23, 31, 34, 35, 36, 38, 41, 179
Colorado Plateau, 134
Colquiri, 149
Columbia River basin, 170
Columbium, 135, 221
Comalco, 174
Combined-cycle system, 74
Combiplant, gas and steam turbine, 101
Comecon, Council for Mutual Economic Aid, 58, 76, 220
Comodoro Rivadavia, 55
Compagnie Française de Pétroles, 62, 220
Compost, 192
Conakry, 174
Concrete, 179
Conference on International Economic Co-operation, 66
Confrontation, 217
Connelsville coal, 33
Consolidated Goldfields, 176
Continental shelf, 68
–, convention of, 5
Continental shield, 11
Conurbation, 26
Cook, Earl, 18
Convergence, Europe-US, 193
Copper, 7, 107, 128, 136, 137, 146, 156, 186, 193, 212–13, 221
–, cable, 165
–, density, 175
–, industry, 210
–, international trade, 145
–, ore, Cyprus, 108
–, refined, 140
–, substitution, 145

Copperbelt, 143
Corning Glass Works, 188
Corundum, 205
Cotta, Bernhard von, 108
Cotton industry, 25
Counter-pressure principle, 84, 99
Coveñas, 54
CRA, Conzinc Riotinto of Australia, 125, 145, 152, 174, 176
Cracking, oil, 45, 136
Creole, 53
Cross-licensing system, 187
Crushed stone, 178
Cryolite, 163, 164, 165
Crypton, 82
Crystal glass, 187
Cullet, 186
Curaçao, 53
Curie, Pierre and Marie, 81
Cuvier, G., 108
Cyprus Mines, 116

Daingerfield, 114
Dampier, 125
Daniel Ludwig, 174
Dannemora, 116
Dar es Salaam, 143
Darling Range, 175
Darwin, 125
Dashkesan, 123
Dasjava, 77
Davy, Sir Humphry, 175
Death Valley, 206
De Beers Consolidated Mines, 204
DCs, developed countries, 207, 219
D-D-reaction, 94
Decentralization, 95
Degtyarsk, 143
Delphi, 168
Depressed districts, 31
Desalination plant, 91
Descartes, René, 107
Deuterium, 82, 94
Devaluation, USdollar, 159
Deville, H.S.-C., 163
Dharan, 64
Diamond, 6, 20, 177, 205
—, alluvial, 204
—, industrial, 203–204
—, jewel, 203–204

Diamond, marine, 204
—, synthetic, 204
Diatomite, 190
Die-casting alloys, 149
Dimension stone, 178
Direct reduction method, 31
Dirty oil, 46
Djebel Onk, 199
Dolomite, 175, 178
Domnarfvet, 221
Donawitz, 119
Donbass, 122
Dortyol, 63
Doomsday prophet, 11, 12
Doubling time, 12, 50
Dow Chemical Co., 175
Dresden, 90, 189
Drogobych, 60
Druzhba, pipeline, 60
Dry lakes, 205, 206
D-T-reaction, 94
Duluth-Superior, 112
Duraluminum, 165
Dzezkazgan, 143
Dzhetygara, 184

Eagle Mountain, 114,
East Liverpool, 189
East St. Louis, 165, 170
Econometric sensitivity analysis, 99, 100
Economic geology, 107
Economic zone, 219
Economy of scale, cement, 180, 183
Ecosystem, 103
ECSC, European Coal and Steel Community, 26, 117
Edjeleh, 66
Eilat, 63
Einstein, Albert, 82, 221
Ekofisk, 56
El Hasa, 200
El-Jadida, 199
El Palito, 53
El Pao, 115
El Romeral, 115
El Salvador, 140
El Teniente, 133, 140
El Tofo, 115
Elbe-Seiten-Kanal, 117
Electric Energy, 97–105

Electric steel furnace, 111, 112
Electricity, 127
—, consumption, selected countries, 102
—, demand, US, 100
—, percentage of total energy consumption, 102
—, price, 103
—, —, US, 98
—, production, 101
—, West Germany, 219
Electric power station, condensation work, 99
—, counter-pressure, 30, 99
Electronic Age, 80
Elf-Erap, 44
Emden, 56
Emerald, 107
Emery, 205
Emission control, car, 163
Energy, 15–105
Energy balance, 70, 79
Energy company, 65
Energy concentrated technique, 221
Energy consumption, Sweden, 18
—, US, 18
Energy crisis, US, 90
Energy policy, 208
—, USSR, 35, 40–41, 219
—, Western Europe, 35
Energy price, 105
Energy saving, 105
Engelhard Minerals & Chemical Corp, 163
Enrichment plant, uranium, 85
Environment, visual blight, 91, 105
Environmental activist, 8, 34, 87, 90, 105
Environmental debate, 48
Environmental deterioration, 104, 156, 221
ERDA, Energy Research and Development Administration (US), 85, 93
Erbium, 221
Erdenet, 133, 145
Erzberg, 119
Es Sider, 66
ESCOM, Electricity Supply Commission (SA), 39
Esmeraldas, 55
Essouira, 199
Euratom, European Atomic Energy Commission, 94
Eurodif, 85
Exchange rate, fixed, 161

Exchange rate, floating, 159
Exploitation credo, 207, 217
Exponential growth, 11–13, 23, 50
Export embargo, Australia, 125
Export-Import-Bank, 220
External economies, cement, 184
Extrapolators of curves, 97
Exxon, former Standard Oil of New Jersey, Esso, 63, 220
Eyde, S., 196

Falconbridge Nickel, 130, 221
Falun, 141, 145
FAMOUS, French-American Mid-Ocean Undersea Study, 108
Faw, 63
FPC, Federal Power Commission, 99, 100
Fermi, Enrico, 80, 82
Ferroalloys, 97
Ferroalloy works, 127
Fertilizer, 192–203, 222
—, consumption, China, 192
—, —, Europe, 193, 194
—, —, India, 192–93, 194
—, —, Japan, 192
—, —, US, 192, 193, 194
Fethiye, 129
Fiat money, 161
Fire clay, 189
Fish, fertilizer, 197
Fischer-Tropsch method, 39, 74
Fissionable fuels, 15
Five-year plan, USSR, 35–36, 122
Flasks, mercury, 156
Flat-glass industry, 186–87
Flin Flon, 140, 150, 162
Float-glass, 187
Floor-price, oil, 209
Flotation enrichment, 137
Fluorine, 10
Flux material, 203
Foggia 168
Food, prices, 222
Forcados, 67
Forrester Team, 103
Fort McMurray, 50
Fortuna-Garsdorf mine, 104–105
Fossil fuels 19, 95, 221
Fourcault, Emile, 186
Fowey, 189

FPC, Federal Power Commission, (US), 99, 100
Framatome, 89
Franklin Furnace, 150
Frasch process, 190, 191
Freeport, 175
Freeport Minerals, former F. Sulphur Co., 130
Frei, Eduardo, 141
Freiberg Mineral Academy, 107
Friedensville, 150
Frejka, Tomas, 13
Friendship pipeline, 60
Frunze, 133
Fúdo-byó, 157
Fuel, 91
–, transport cost, 91
Fuel prices, 103
Fuel recovery plants, 86, 87
Fuller's earth, 190
Fusion power, 15, 18, 87, 93
Fusion reactor, 94
Futures, 137
Gabes, 199
Gadolin, John, 221
Gadolinium, 221
Gafsa, 199
Gagnon, 114
Galena, 154
Gällivare, 116, 117
Galvanization, 149
Gas, see also natural gas, 72, 77
Gas deals, 74
Gas diffusion plant, 85
Gas turbine generator, 101
Gasolene, 72
Gazli, 77
GCFBR, Gas cooled fast breeder reactor, 93
Gdansk, 202
Gécomines, Générale Congolaise des Mines, 142, 221
Gelsenberg, 44
Gem, 177
General Electric, 84, 85, 139
Geometric progression, 11
Geophysical exploration, 43, 53
Geopolitical embargo, 125
Geopolitics, 65, 208
George, Henry, 11
Geothermal energy, 15, 18, 48, 102
German silver, 137, 161

Germanium, 221
Getty, 44
Ghawar, 64
Gladstone, 174
Glass, 73, 166, 185–88
–, international trade, 188
–, polishing, 136
Glass container, 187
Glassware, international trade, 188
Glomar Explorer, 128
Globe, 137
Goa, 124, 127
Gold, 6, 107, 131, 157, 158, 186, 221
–, auction, 159
–, demonetized, 159
–, free market, 159
–, ingot, 158
–, native, 158
–, placer, 158
–, price of, 163
–, reserve, 159
–, world production, 160
Gold reef, 159
Gold share, 159
Gora Kachkanar, 134
Gorky, 61
Gove Peninsula, 175
Grain purchases, Soviet, 222
Grand Coulee Dam, 170
Gränges, 116, 211
Grängesberg, 116, 117
Granite, 178
Graphite, 20, 177, 190
Gravel, 185
Gravimeter, 44
Gray Canyon, 179
Great Lakes, US, 179
Great Salt Lake, 201
Green manure, 192
Green Mountains, 179
Green Revolution, 192, 197
Gresham's law, 161
Griesheim Elektron, 175
Groningen, 75
Growth rate zero, 12
Groznyj, 57
Guano, 198, 200
Guggenheim brothers, 140
Guggenheim process, 196
Gulbenkian, C.S., 220

Guleman, 129
Gulf, 53, 64, 209
Gulf area, VII, 61–66
Guryev, 60
Guyana Shield, 171

Haber, Fritz, 196
Haber-Bosch-method, 196
Hahn, O., 82
Haifa, 62, 63
Hall, C.M., 164, 165
Hammer, Armand, 220, 221
Harjavalta, 146
Hartford-Empire, 187, 222
Harz, 201
Hassi Messaoud, 66
Heavy hydrogen, 94
Heavy metals, 156
Heavy water, 94, 97
Heavy water reactor, 90
Hematite, 109
Héroult, Paul, 164, 165
High-voltage transmission, 105
Hiroshima, 80
Hitler, Adolf, 127
Hoarding, 161
Högbom, Ivar, IX
Hollard Street, 159
Holmium, 221
Homestake mine, 161
Hoover, Herbert, 107
Huanuni, 149
Hughes, Howard, 128
Hunt, A.E., 164
Hurricane Creek, 165
Hüttenberg, 119
Hutton, J., 108
Hydraulic cement, 180
Hydroelectric power station, 95
Hydrocarbons, 41, 70, 79
Hydrogen, 10
Hydrogen bomb, 82
Hydrogen fusion, 82, 95
Hydrological cycle, 15, 95, 192

Idemitsu Kosan, 44
IEA, International Energy Agency, 209, 220
IG Farbenindustrie, 175
Ilmen Range, 177
IGU, International Geographical Union, 208

Ilmenite, 176
–, world production, 177
Imatra, 146
IMF, International Monetary Fund, 159
Import quotas, US, 208
Import substitution, 197
Inco, International Nickel Company, 130, 221
Indian Point, 90
Indigirka River, 160
Industrial complexes, 41
Insulation of houses, 105
Iodine, 195
IPC, Iraq Petroleum Co., 63, 220
Irkutsk, 168
Iron, 193
–, density, 175
Iron and steel industry, 31, 210
–, location, 111
Iron ore, 108–126, 177, 178, 180
–, Australia, 211
–, Chile, 211
–, imports US, 112
–, US, 211
Irrigation, 221
Itabira, 115

Jarrahdale, 175
Jashno-Chingan, 127
Jerevan, 168
Jos Plateau, 149
Jussarö, 117

Kainite, 201
Kaiser, 166, 174
Kaldo-converter 110, 221
Kalgoorlie, 131
Kalling, B., 221
Kamensk-Uralski, 168
Kandalaksha, 168
Kaolin, 188
Karaganda, 123
Karat, 158
Kasekelesa, 128
Katanga, see also Shaba, 128, 135, 221
Kennecott, 138
Kerch Peninsula, 122
Kerosene, 72
Keweenaw Peninsula, 138
Kharg Island, 62, 65, 192
Khavdarken, 156

Khor al Amaya, 63
Khouribga, 199
Kidd Creek, 150
Kildonan, 129
Kimberlite, 203
Kimberley, 150, 154, 162, 204
Kinetic energy, 94, 95
King Idris, 66
Kipushi, 152
Kirishi, 61
Kirkenes, 117
Kirkland, 161
Kirkuk, 62
Kirovobad, 168
Kirovsk, 168, 198
Kiruna, 116, 117
Kisenga, 128
Kislaya Guba, 95
Kitimat, 171
KMA, Kursk Magnetic Anomaly, 122
Knob Lake, 114
Knoxville, 179
Kohtla-Jarve, 77
Kokkola, 146
Kolari, 117
Kolyma River, 160
Konstantinovka, 152
Kostamus, 117, 123
Kostomukscha, 123
Kounradskiy, 143
Krasnaja Chapochka, 168
Krasnodar, 77
Krasnojarsk, 168
Krasnooktjabrski, 168
Krasnoturinsk, 168
Kremenchug, 122
Kristiansand, 130, 146, 221
Krivoj Rog, 120, 122
Krugerrand, 159
Krupp-Renn process, 125
Krypton, 84
Kufra, 66, 21
Kukes, 129
Kuruman, 128
Kuzbass, 122
Kwinana, 175

La Goulette, 199
La Salina, 53
La Skhirra, 66

Labrador, 112
Labrador Iron Through, 114
Lago Agrio, 55
Laissez-faire-policy, 70, 208, 209, 210
Lake Athabasca, 50
Lake Maracaibo, 52, 53
Lamco, Liberian American-Swedish Minerals Co., 117, 126, 211
Lancashire, 25
Land rent, VIII, 217, 220, 222
Lannon, 178
Lapis lazuli, 107
Lapland, 198
Larder Lake, 161
Las Vegas, 176
Laser fusion, 93
Laurion, 107, 153, 158
Law of the Sea, Conference, 5, 219
LD-converter, 110, 111, 119, 221
LDCs, less developed countries, 139, 207, 212, 219
Le Creusot, 130
Lead, 7, 134, 153, 156, 221
—, smelter production, 155
Lead coins, 153
Lead ore, mining, 153
Lead smelters, location, 155
Leakey, L. B. S., 107
Leduc, 49
Lehigh Valley, 180, 181
Lenin, V.I., 220
Leninogorsk, 143, 153
Les Baux, 164, 167
Leslie, D.C., 83, 86
Leuna works, 201
Levy, Walter J., 209
Lewis, John, L, 219
Libbey-Owens-Ford, 187
Liebig, Justus, 193
Lignite, 20, 26, 34
Lignite briquettes, 23, 30
Light metals, 163–177
Light-water reactor, 15, 82, 86
—, safety of, 104
Lilienthal, David, 85
Lime, 177, 178, 203
Limestone, 11, 178, 179, 180, 203
Limoges, 189
Limonite, 109
Linz, 119

Lipetsk, 120, 122, 123
Listerhill, 166
Lithium, 94
Litosphere, 10, 11
Lovozero, 135
LKAB, Luossavaara Kirunavaara AB, 116
LME, London Metal Exchange, 137, 213, 222, 223
LMFBR, Liquid metal (cooled) fast breeder reactor, 93
LNG, liquefied natural gas, 72, 74, 76
LNG-tanker, 66, 71
Lobito, 143
Loco, 137, 213
Locomotive, wood-fired, 17
Lode, gold, 158
–, tin, 147
Logistic curve, VIII, 11–13, 17, 50
Lökken, 141
Long-term capital, interest rate, 96, 97
Long-term contract, 211
Longview, 170
Lorraine, 117, 118
Los Alamos, 80
LPG, liquefied petroleum gas, 72
LPG-tanker, 71
Luleå, 116
Lurgisystem, 32
Lutetium, 135
Lydenburg, 163
Lynn Lake, 130

Macapá, 128
MacKenzie, R., 75
Macro-nutrients, 193
Magdeburg, 151
Maghreb, 125, 199
Magnesite, 175
Magnesium, 10, 97, 175–76, 193
–, density, 175
–, US production, 175
–, world production, 176
Magnetite, 109
–, titaniferous, 125
Magnetometer, airborne, 7, 44
Magnitogorsk, 122, 123
Magnox, 88
Majchura, 131
Majkop, 57
Malm, 117

Malmexport, AB, 117
Malthus, Thomas, 11, 12
Mamonal, 54
Manganese, 7, 10, 127, 193
–, world production, 128
Manganese nodules, VIII, 108, 128
Manhattan Project, 82
Manhattan, tanker, 48
Mano River, 126
Mansfeld, 145, 201
Manufacturing Belt, 47, 73
Manure, 192
Marathon, 44
Marble, 178, 179
Marcona mine, 116
Marine evaporites, 201
Market location, cement mill, 180
–, glass, 187
Marl, 180
Marmora mine, 114
Marsa el Brega, 66, 76
Marsa el Hariga 66
Marsh gas, 72
Marshall, A., 184
Martin, Pierre, 186
Maruzen, 44
Marx, Karl, 207, 214, 217
Masjid-i-Sulaiman, 61
Mass deficiency, 81, 82
Mass number, 81
Massena, 170
Material concentrated technique, 221
Mattagami, 150
MCs, Multinational corporations, 97, 207
Megalopolis, 180
Meissen, 189
Meitner, Lise, 82
Mena al Ahmadi, 65
Mene Grande, 53
Mercury, 7, 156–57
–, catastrophy, 157
–, disinfecting of seed, 157
–, world production, 157
Mesabi range, 111
Metal Bulletin, 160
Metal Marketing Corporation, 143
Metals, major, production, 126
–, structural, 175
Metamorphic rocks, 11
Methane, 72

Metz, 117
MHD, Magnetohydrodynamic generator, 102
Mica, 177
Michipicoten, 114
Micro-nutrients, 193
Middleburg, 134
Middlesbrough, 119
Midland, 175
Mikhailov, 122
Minamata, 157
Mine gas, 72
Mineral fuel, 19
Mineral imports, Japan, 126, 215
–, US, 213
Mineral production, 221
Mineral realm, VII, 19
Mineral reserves, 111
Minerals, VIII
–, sea water, 175
–, international trade, 9, 97, 107, 216, 207–217
Minette ores, 117
Mini-steelworks, 112
Mirnyj, 204
Misch metal, 135, 136
Mo i Rana, 117
Mobil, 63, 220
Mobile, 170
Mohorovičić discontinuity, 108
Molybdenite, 133
Molybdenum, 127, 133, 193
–, world production, 133
Mombasa, 143
Monazite sand, 92, 136, 176
Mond, Ludwig, 130
Mond Nickel, 130, 221
Monetary standard, 157
Monnet, Jean, 26, 117
Monopoly pricing, OPEC, 208, 222
Montana-Wyoming, coal, 8, 39
Monte Amiata, 156
Montreal, 49
Moral dilemma, 221
Morocco, Sultan of, 222
Mosander, C.G., 221
Mother lode, 203
Mouanda, 128
Mountain Pass, 136
Mount, T., 100
Mt. Enid, 125

Mt. Goldsworthy, 125
Mt. Isa, 145, 154
Mt. Newman, 125
Mt. Tom Price, 125
Mt. Whaleback, 125
Mozyr, 60
Mud, for drilling, 189
Mulhouse, 201
Murmansk, 123, 220

Nadvoitsy, 168
Nagasaki, 80
Nancy, 117
Narvik, 116, 117
National Iranian Oil Co., 62
National motorways, 31
Nationalization, mines, 212
Natural cement, 179
Natural gas, 17, 18, 31, 32, 34, 41, 62, 70–79, 98, 166, 186, 187, 196, 212, 219, 220
–, aluminum, 171
NCCM, Nchanga Consolidated Copper Mines, 143
Nchanga, 143
Neolithic man, 107
Nepheline, 168
Neptunist school, 107, 108
Net reproduction rate, 13, 99, 100
Neutrons, 81
New Broken Hill, 145, 152
New Deal, US, 197
New Kensington, 166
Newcastle, 125
Niagara Falls, 170
Newcomen, Th., 23
Nickel, 6, 10, 127, 128, 129–31, 136
–, international trade, 131
–, oxide (laterite) ores, 130, 131
–, sulfide ores, 130, 131
–, world production, 130
Night soil, 192
Nikopol, 127
Niksic, 168
Nimba, 126
Nimonic, 135
Niobium, 135
Nitric acid, 196
Nitrogen, 10, 192, 193, 194–197, 222
–, fertilizer, production, 195
Nixon, Richard M., 158

Nizhniy Tagil, 163
Nodules, VIII
Non-ferrous metal ores, international trade, 146
Nonmetallic minerals, 177—192
Non-renewable materials, 12
Noranda-Rouyn, 140
Norilsk, 130, 143 163
Norsk Hydro, 196
North Broken Hill, 152
North Sea, oil, 44, 56—57
North Staffordshire, 25
Norwegian saltpeter, 196
Notodden, 196
Nouadhibou, 126
Novokuznetsk, 168
Novorossisk, 57, 61
NRC, Nuclear Regulatory Commission, US, 85
Nsuta, 128
Nuclear energy, 15, 17, 80—94, 102, 219
Nuclear fission, 82, 84
Nuclear mining, 137
Nuclear physics, 86, 87
Nuclear weapons, 87
Nucleons 81

Oak Ridge, 85, 93
OBM, Oxygen Boden Maximilianhütte, 111
Obninsk, 83
Occidental, 44, 55, 66, 220
OCP, Office Chérifien de Phosphates, 199
Odell, Peter, 55, 209
OECD, Organization for Economic Cooperation and Development, 219
Off-shore, continental shelves, 65, 68
—, drilling, 55, 57
—, minerals, 5—6
—, oil, 47, 105
—, prospecting, 69
Oil, 17, 30, 34, 41, 42, 43—70, 72, 73, 86, 221, 222
—, boom, VII
—, East Asia, shelf potential, 214
—, exports, China, 214
—, import quotas, US, 210
—, imports, Japan, 214
—, off-shore, 5—6
—, production costs, 54, 208, 209, 210, 220, 222
—, Venezuela, 210

Oil cartel, 54
Oil crisis, 87, 90, 97, 99, 210, 212, 220
Oil cycle, energy, 86
Oil equivalents, 72
Oil refineries, 45, 46
Oil sand, 18
Oil shale, 18, 70, 77, 79
Oil strategy, North Sea, 214
Oil trap, 42, 47
Oilstone, 205
Okan, 67
Olenegorsk, 123
On-shore, oil, 57
Oolithic limonite, 122
Oolithic siderite, 109
OPEC, Organization of Petroleum Exporting Countries, 35, 54, 62, 66, 99, 208, 209, 214, 217, 220, 222
Opencuts, see also open-pit mining, 111, 118
Open-hearth furnace, 110, 112, 186, 198
Open-pit mining, 31, 34, 104, 171
‚Oppenheimer Empire', 160, 204
Orapa, 204
Ore, metal content of, 9
—, transportation system, 112
Ore carriers, Great Lakes, 112
Orenburg, 77, 220
Ordzhonikidze, 152
Oron, 200
Ørsted, H.C., 163
Osmium, 162
Otanmäki, 117, 134, 177
Outukumpu, 135
Outukumpu Company, 146
Owens, M.J., 187
Owens-Illinois, 187, 222
Oxelösund, 116
Oxygen, 10
Oxygen process, steel, 112
Oyster shell, 180

Paducah, 85
Palladium, 162, 163,
Panadjuriste, 146
Panama Canal, 48, 140
Panasqueira, 132
Paraguaná Peninsula, 53
Paris Exhibition 1854, 163
Parsees, 43
Pavlodar, 168

Index 243

Pazardzik, 146
Peat, 20
Pechenga, 130
Péchiney, 166, 168, 174
Peekskill, 205
Pellets, iron ore, 109, 189
Pemex, Petróleos Mexicanos, 44, 52
Penang, 147
Periodic system, 81
Permalloy, 129
Persian Gulf, see the Gulf
Perth Amboy, 140
Peter the Great, Czar, 120
Petrobrás, 44
Petrochemical complex, 77
Petrofina, 44
Petroleum, 41–79
–, production cost, 210
–, sour, 190
Petroleum complexes, 41
Petroperu, 55
Petroven, 54
Petsamo, 130, 221
pH-value, 203
Phelps Dodge, 138
Phénix, 93
Philadelphia, 48, 114
Phillips, 44
Phosphate, 168, 190
Phosphorus, in iron, 109
Phosphorus, 10, 193
Phosphorus cycle, 192
Phosphorus pentoxide, 197
Photosynthesis, 17, 95, 194
Pierrelatte, 85
Pig iron, international trade, 119
–, production, 112
Pikaleva, 168
Pilbara, 7, 111, 211
Pilkington Brothers, 186
Pipes, 203, 204
Pipeline, coal, 40
–, system, 60, 70, 73, 77
Pirdop, 146
Pitchblende, 81, 82
Pittsburgh, 166
Pittsburgh Reduction Company, 164
Pittsburgh seam, 21
Placer mining, 147, 158
Plant breeding, 195

Plasma, 94
Platinum, 162–163
Platinum catalysts, 163
Platinum group, 157
Plesetsk, 168
Plock, 61
Ploesti, 55, 56
Plutonist school, 108
Plutonium, 86, 104
Plutonium bomb, 80
Pocahontas coal, 33
Poe Lock, 112
Pointe Noire, 128
Political geography, 208
Pollution, 156
Pollution abatement, cement, 180
Pompeji, 153
Ponta do Tubarão, 115
Population increase, world, 99
Porcelain, 188–89
Pori, 146
Port Hedland, 125
Port Kembla, 125
Port Lambert, 125
Port Nolloth, 204
Port Radium, 162
Portland cement, 179–184, 222
Portsmouth, 85
Postindustrial society, 13
Postmasburg, 125, 128
Potash, 201–02
–, world production, 202
Potassium, 10, 193, 201
Potassium cycle, 192
Pottasium oxide, 201
Potosi, 133
Potrerillos, 140, 141
Potteries, 25, 189
Power, political, 208, 222
PPG Industries, former Pittsburgh Plate Glass Co., 186, 187
Precious metals, 157–63
–, state secret, 160
Primary energy, 97, 102
Producer-gas, 17
Propane, 72
Protein, 194
Protoplasm, 194
Prudhoe Bay, 48, 75
Portland, 49

Power gas, 32
Puerto Miranda, 53
Puerto Ordaz, 115
PUK, Péchiney Ugine Kuhlmann, 166
Pumice, 205
Punta Cardón, 53
Puzzolan, 179
PWR, Pressurized Water Reactor, 84, 89, 90, 92, 104
Pyrites, 191

Qatif, 64
Qinhuangdao, 68
Quartz crystal, 177
Quartz vein, 158
Quicksilver, 156
Quincy, 178

Raajärvi, 117
Radiation, 104
Radiation shield, 153
Radioactive fallout, 82
Radioactive waste, final storage, 86, 104
Radioactivity, 81
Radio-therapy, 135
Rance estuary, 95
Rand, 159
Raniganj, 123
Ranstad, 82
Ras Khafji, 64
Ras Lanuf, 66
Ras Tanurah, 64, 65
Rasdan, 168
Rasmussen report, 86, 104
Rautaruukki, 117
Raw materials, VII
Raw phosphate, 197–200
–, prices, 199
–, world production, 198
Razdolinsk, 133
RCM, Roan Consolidated Mines, 143
Reactor hearth, 84
Rechitsa, 60
Recovery rate, 154
Recycling, 137
Red Lake, 160
Reforma-field, 52
Refuse, 198
Reinforced concrete, 178
REM, Rare-earth metal, 135, 221

Remote sensing, 126
Renault, 207
Renewable materials, 12
REO, Rare earth oxides, 135, 136
Reserves, 5–11, 221
Resources, 5–11, 15, 221
Rheinische Braunkohlenwerke, 104–105
Rhizobium, 194
Reynolds, 166
Rhodium, 162
Rio Blanco, 141
Rio Tinto, 145
River steamer, wood-fired, 17
Rjukan, 196
Rockefeller, John D., 220
Rogers City, 179
Rönnskär, 146
Room-and-pillar system, 33, 39
Roosevelt, Franklin D., 158
Röros, 141
Rotary kiln, 180, 222
Rotschild interests, 130
Rottenstone, 205
Rotterdam, 48, 206
RTZ, Rio Tinto-Zinc Corp, 152
Rubidium, 10
Ruby, 205
Ruhr, 118
Rumaila, 63
Ruseifa, 200
Russian oil, 210
Rustavi, 123
Rustenburg, 129, 163
Ruthenium, 162
Rutherford, Ernest, 81
Rutile, 176

Safaniyah, 64
Safi, 199
Saiansk, 170
St. Helens, 186
St. Lawrence Seaway, 114
Saldanha, 126
Saline country, 170
Salt, 177
Salt dome, 47, 191
Salzgitter, 117, 118
San Juan, 116
San Luis Potosí, 133
Sand, 177, 185

Sandblasting, 205
Sandstone, 11, 178, 179
Sanford Lake, 177
Santander, 120
Sapphire, 62, 205
Sarany, 129
Saratov, 77
Sarir, 67
Sarnia, 49
SASOL, South African Coal, Oil and Gas Corp., 39
Scandium, 135, 221
Scarce metals, 156
Scheelite, 131
Schefferville, 114
Schmehausen, 93
Schneider, 130
Schwedt, 61
Scrap-based metal, energy, 111, 112
S-curve, 12
SDR, special drawing rights, 161
Sea bed, minerals, VIII, 5
Sea currents, 95
Sea water, minerals, 5
Searles Lake, 206
Secondary energy, 97, 102
Selenium, 156
Sedimentary rocks, 11
Selukwe, 129
Semilogarithmic diagram, 50
Sensitivity, 103
Sept Iles, 114
Seven Islands, 114
Seven Sisters, 44, 210
Sfax, 199
SGHWR, Steam Generating Heavy Water Reactor, 89
Shaba, 128, 221
Shale, 11, 179, 180, 221
Shawinigan Falls, 170
Shell, Royal Dutch/Shell, 62, 63, 67, 174, 220
Sherman, Antitrust Act, 152, 220
Shevchenko, 60
Shippingport, 90
Sibenik, 168
Siderite, 109
Sidon, 64
Siemens, William, 186
Siemens-Martin-furnace, 110, 186
Sierra Club, 34

Sierra de Bahoruco, 174
Sierra Gorda, 141
Signal Hill, 45
Silica sand, 180
Silicon, 10, 127
Silicon carbide, 205
silk industry, 31
Silver, 7, 154, 157, 161–62, 221
—, byproduct, 162
—, standard of value, 161
—, world production, 162
Silver lobby, 152
Single supplier, risk, 125
Sintering, 109
Sishen, 126
SITC, Standard International Trade Classification, 213
Sjebelinka, 77
Slag, 180
Slate, 177, 178, 179
Sleeping partner, 65
SLN, La Société Le Nickel, 130
Slochteren, 75
Smålands Taberg, 177
Smidth, F.L., 222
SNG, synthetic natural gas, syn-gas, 74
Société Général, 142
Société Général des Minerais, 142
Soda, 168
Sodium, 10
Soil conditioner, 20
Solar energy, 15, 17, 18, 87, 93
Solar evaporation, 201
Solder, 146, 147
Soligorsk, 202
Solikamsk, 202
Solvay method, 164
Soo Canal, 112, 211
Sparrows Point, 115
Special glass, 187
Speer, Albert, 127
Split, 168
Spokane, 170
Spot market, 138, 210, 222
Stakhanov, Alexei, 219
Stalin, Josef, 80
Standard Oil, 209, 220
Standard Oil Indiana, 62
Standard Oil of California, 63

Standard Oil of New Jersey, see also Esso, Exxon, 53
Stary Oskol, 122
Stassfurt, 201
Stavanger, 56
Stavropol, 77
Sté Alsthom, 89
Steam engine, 23, 95
Steam turbine, 101
Steel, international trade, 124
—, production, 118, 120
Steel furnace, 127
Steel industry, 35
Steep Rock, 114
Stockpiles, 214
Stoke-on-Trent, 189
Stone, 177, 178—79
Strait of Hormuz, 61
Strassmann, F., 82
Strategic materials, 127, 136, 214
—, Germany WW II, 127
Strip mine, see also open-pit mining, 33, 35, 36, 39, 122
Strontium, 10
Sudbury, 130, 135, 140, 163, 221
Sudr, 66
Sugar beet tops, 201
Sulfide ores, 158
Sulfur 10, 109, 177, 190—92, 193
Sulfur effluvia, 72, 86, 138, 191
Sulfur recovery, 191
Sulfuric acid, 190
Sulitjelma, 146
Sumgait, 168
Sun flower stalks, 201
Sunshine, 162
Super-Phénix, 93
Superphosphate, 190, 200
Supreme Court, US, 166
Surigao, 124
Svappavaara, 117
Sylvite, 201
Synthetic gas, 32, 34, 41
Syracuse, 189
Szazhalombatta, 61

Taching, 68
Tacoma, 170
Taconite ore, 109, 112, 211
Talara, 55, 200

Tampico, 50
Tantalum, 135, 221
Tantalum carbide, 205
Tapline, Trans-Arabian Pipeline, 64
Tar sand, 50, 53, 70, 105
Tata Works, 123
Tees-side, 56
Techsnabexport, 136
Terbium, 221
Territorial water, 219
Test pile, 91
Tetyukhe, 152
Texaco, 63
Thabazimbi, 125
Thames, River, 181, 184
Thermal energy, 95
Thermal neutrons, 92
Thermal pollution, 103
Thermic reactor, 103
Thermodynamics, laws of, 99
Thermo-electric power plant, 34, 84
Thermonuclear reactor, 82
Thetford Mines, 184
Thierry, 142
Thomas-Gilchrist, converter, 110, 116, 117, 198
Thompson, 130
Thomson, J. J., 81
Thorium, 17, 18, 82, 92, 105, 136
Thulium, 221
Tidewater energy, 15, 18, 95
Timmins, 161, 162
Tin, 6, 7, 136, 146, 221
—, international trade, 149
—, world production, 147
Tin oxide, 147
Titanium, 7, 10, 97, 133, 136, 169, 176—77
—, density, 175
Titanium boride, 206
Titanium carbide, 205
Titusville, 47, 55
Togliatti, 77, 220
Tokmak, 127
Tokomak, 94
Toquepala, 142
Trace elements, 193
Trans-Alaskan Pipeline, 8
Traprock, 178
Trenton, 189
Tricastin, 85

Tripoli, 62, 63, 205
Tritium, 82, 94
Trombetas River, 174
Tropical countries, water power, 97
Troutdale, 170
Troy system, 158
Troyes, 158
Truman, Harry, 80, 85
Tuapse, 57
Tula, 120, 122, 123
Tungsten, 7, 127, 133, 221
–, world production, 132
Tungsten boride, 206
Tungsten carbide, 205
Turgay, 133
Turner, Valley, 49
Turn-key plant, 89
Turquoise, 107
TVA, Tennesee Valley Authority, 85, 171
Type metal, 147
Tyrnyauz, 131, 133
Tyrrell, T., 100
Tyumen, 58, 61, 77

UKC, Urals-Kuzbass Combine, 122, 123
Ulu-Teljak, 127
UMW, United Mine Workers, 219
Uniformism, doctrine of, 108
Union Minière du Haut-Katanga, 142, 221
Uranium, 7, 17, 18, 79, 80, 81, 82, 86, 91, 92, 104, 105, 126, 221
–, Canada, 134
–, South Africa, 159
–, Zaire, 134
Uranium bomb, 82
Uranium cycle, energy, 86
Uranium mine, risks, 104
Uranium 235, 82, 85
Urea, 196
Urenco, 85
Urucum, 128
US Steel Corp., 115, 128, 211
Usa, 127
Ust-Kamenogorsk, 152
Utah Construction and Mining, 116
Uzhgorod, 220

Val d'Or, 184
Valdez, 48
Vanadinite, 134

Vanadium, 7, 10, 127, 134
–, world production, 134
Vancouver, 170, 191
VEBA, 44
Vein, 147
–, gold, 158
Velasco, 175
Venezian, 153
Ventspils, 58, 61
Victoria gold rush, 158
Vitória, 115
Visual blight, 91, 105
Volcanik necks, 203
Volgograd, 168, 170
Volkhov, 168
Volta Redonda, 115
Volta-project, 97
Vorkuta, 123
Vraca, 146
Vylder, Stefan de, 222, 223

Wabana, 114
Wabush Mines, 114
Warrior Basin, 112
Waste heat, 84, 99, 103–104
–, recovery, 103
Water, fresh, VIII–IX
Water power, 15, 17, 37, 94–97, 102, 166, 212, 219
–, aluminum, 171
Water wheel, 18, 94
Watt, James, 23
Waves of the sea, 95
Wedgwood china, 189
Weipa, 174
Welkom-Odendaalsrus, 159
Werner, A. G., 107
Western Electric, 139
Westernport, 69
Westinghouse, 84, 85, 89, 139, 166
White Pine, 138
Whyalla, 125
Wilmington, 206
Witwatersrand, 159
Wolfram, 131, 133
Wolframite, 131
Wood, 17, 18, 95, 221
Workshop of the world, 95

Xenophon, 107
X-rays, 80

X-strahlen, 80

Yampi Sound, 125
Yaroslavl, 61
Yellowknife, 160
Yom Kippur War, 65
Youssoufia, 199
Ytterbium, 221
Ytterby, 221
Yttrium, 135

Zàjecar, 146
Zagros Mountains, 61
Zambales, 129

Zaporozhye, 168
Zechstein, 202
Zelten, 66
Zero population growth, 13, 100, 221
Zimmermann, Erich W, VII
Zinc, 7, 10, 97, 136, 149–152, 154, 156, 193, 221
–, world production, 150
Zinc Corporation, 152
Zink, sulfide ore, 153
Zirconium, 10, 127, 136
Zoroastrians, 43
Zouerate, 126
Zubair, 63
Zueifina, 66